Abhik Narayan Choudhury

# Quantitative phase-field model for phase transformations in multi-component alloys

Schriftenreihe
des Instituts für Angewandte Materialien
*Band 21*

Karlsruher Institut für Technologie (KIT)
Institut für Angewandte Materialien (IAM)

# Quantitative phase-field model for phase transformations in multi-component alloys

by
Abhik Narayan Choudhury

Dissertation, Karlsruher Institut für Technologie (KIT)
Fakultät für Maschinenbau
Tag der mündlichen Prüfung: 17. Februar 2012


**Impressum**

Karlsruher Institut für Technologie (KIT)
KIT Scientific Publishing
Straße am Forum 2
D-76131 Karlsruhe
www.ksp.kit.edu

KIT – Universität des Landes Baden-Württemberg und
nationales Forschungszentrum in der Helmholtz-Gemeinschaft




KIT Scientific Publishing 2013
Print on Demand

ISSN 2192-9963
ISBN 978-3-7315-0020-9

# *Quantitative phase-field model for phase transformations in multi-component alloys*

Zur Erlangung des akademischen Grades

**Doktor der Ingenieurwissenschaften**

der Fakultät für Maschinenbau

**Karlsruher Institut für Technologie (KIT)**

genehmigte

**Dissertation**

von

**MTech. Abhik Narayan Choudhury**

Tag der mündlichen Prüfung: $17^{th}$ February 2012
Hauptreferent: Prof. Dr. rer. nat. Britta Nestler
Korrereferent: Assosciate Prof. Dr. Habil. Mathis Plapp

*I thank my family and especially my parents for their constant support and encouragement which they provided whenever I was in need. I thank Prof. Britta Nestler for the support and guidance during my PhD. Special thanks to Mathis Plapp for giving periodic technical feedback on the developments and most of all the working group here, who provided a congenial working environment with lively discussions and technical support*

# Contents

Contents

# List of Figures

# List of Tables

# Chapter 1

# Motivation and outline

In the present doctoral thesis the development of a quantitative phase-field model for modeling phase transitions in multi-component sytems is presented. The work is categorized into four main sub-divisions as,

- Exemplary applications of the phase-field method

- Modifications to an existing phase-field model

- Asymptotics

- Validation

To start with, investigations are performed with the aim of modeling some essential features in the process of solidification using the phase-field model of H.Garcke and B.Nestler and others [79], for dendritic growth, binary eutectic, peritectic and ternary eutectic systems. Apart from being of scientific importance, aiding in the understanding of the physics of solidification in various phenomena, the studies lay a foundation of basic knowledge required for the development of a quantitative phase-field model, highlighting the challenges to be overcome in the development of an effective model. Following, is the general overview of the overall work.

## 1.1. Exemplary applications of the phase-field method

Two studies of interest are presented: (I) Growth in the (Fe-C) peritectic system, of $\delta$ (ferrite(pro-peritectic phase)) and the $\gamma$ (austenite(peritectic phase)) [19]. The free energy of the phases were modeled using the ideal solution model such that the liquidus and solidus slopes along with the concentrations of the respective phases at the peritectic temperature fit to the actual phase diagram. The model parameters related to the interface width and surface excesses were adjusted to derive the surface energy of the different interfaces presented in literature or the intended value to be set in the simulations. Although the surface energies of the solid-liquid interfaces are known quite accurately, the solid-solid surface energies $\tilde{\sigma}_{\alpha\delta}$ are unknown. In this study a range for the solid-solid surface energies is derived on the basis of the occurence of the engulfing morphology(pro-peritectic phase engulfing the peritectic phase above the peritectic temperature, and vice-versa below the peritectic temperature.) It is noticed that the solid-solid surface energies, strongly influence this growth morphology at given supersaturations, enabling the isolation of such a range in the solid-solid surface energies. In addition, critical nuclei are numerically calculated through the solution of the Euler-Lagrange equations comprising of the stationary phase-field and the concentration equations. Through this, the homogeneous barrier to nucleation is computed for the nucleation of the $\delta$ and $\gamma$ phases in the liquid at various compositions. As the Fe- concentration increases, it becomes more favorable to nucleate both phases. Simulations of nucleation events are performed with stochastic noise coupled with dendritic growth of the pro-peritectic phase and the sites of nucleations in the non-uniform concentration field, confirm to the predictions derived from the calculations of the barrier to nucleation.

(II) The second example used for the investigation of solidification is the study of ternary eutectics [20]. In particular, special attention is devoted to configurations during thin-film growth. In contrast to binary eutectics where the only possibility is $\alpha\beta\ldots$, there exists a number of possibilities for the growth patterns of ternary eutectics in thin-film growth, eg $\alpha\beta\gamma\ldots$ $\alpha\beta\alpha\gamma\ldots$. A theoretical study is performed of the Jackson-Hunt type for the various configurations, resulting in expressions of the undercooling as functions of lamellae spacing for given velocities. Corresponding comparisons were made with simulations for a symmetric ternary eutectic system modeled using ideal free energies. Good aggreement was achieved between the simulations and theory in the prediction of the spacings at minimum undercooling and the undercoolings themselves. For large spacings, the lamellae exhibit oscillatory instabilities. The symmetry elements present in the different configurations match those of the underlying symmetry elements of the configuration. Some of these symmetries match those found previously for binary eutectics. Although some of the symmetry modes are pertinent with respect to the specially constructed, symmetric phase diagram, their occurence in real alloys cannot be ruled out without an examination. An additional instability that was found for configurations other than the simplest configuration $\alpha\beta\gamma$, is that below a particular lamella spacing, the lamella are unstable towards elimination. The critical spacing below which this instability occurs, can be well explained on the basis of the theoretical calculations. Furthermore, simulations of directional solidification of bulk samples in three dimensions is also performed.

Two types of morphologies were isolated depending on the concentration and the symmetry of the phase diagram. For the symmetric phase diagram at the eutectic temperature, a hexagonal pattern was achieved in the cross-sectional view, starting from a random configuration. For a slightly asymmetric phase diagram a semi-regular brick structure is obtained which is also observed in experiments. While this does not cover all the possibilities, it certainly presents an outlook into the variety of structures.

## 1.2. Model modification

On the basis of these two studies, there were two conclusions with respect to the applicability of the model for the case of phase transformation in real alloys. With regards to the length scale that can be simulated, it was noticed that there exists considerable limitations to the grid resolution that can be used, and in most cases this presents a significant computational overhead. It has been well established in literature, that when the free energies are interpolated in the form used in the WBM type models[134], there exists an additional length-scale coming from the variation of the grand potential excess across the interface [15, 55, 77, 114]. For systems, in which this term becomes largely dominant, the surface energy and the interface thickness loose their independence (surface energy and interface thickness are usually, two independent parameters in simulations). This limits, the interface widths to smaller values, and hence the domains that can be simulated. Consequently, this results in a limitation in the physical sense, because the processing conditions that can be simulated with such an approach gets narrower. Additionally, one must pre-calculate the contribution of the chemical free energy excess to the surface energy, in order to choose the right simulation parameters. On a more technical note, it is also diffcult to perform the thin-interface asymptotic analysis (described later) for such a model, because, the equilibrium properties scale with the interface thickness. This challenge motivates a change in the modeling idealogy with the following aims,

- Construction of a model with efficient flexibility in choosing the right parameters,

- Easy applicability and extendability to any alloy system,

- Performing of thin-interface asymptotics with universally applicable results.

In literature, there exists two types of modeling idealogies for deriving a phase-field model, where the equilibrium properties such as the surface energies are independent of the free energies of the respective phases. In the first type, the functional argument is just the concentration field [30], while in the other [55], the arguments are the concentration fields corresponding to respective phases in the system, eg: $c_\alpha, c_\beta$ etc. In the case where a single concentration field is used, the free energy contribution is decomposed into enthalpic and entropic contributions which are interpolated independently, such that at equilibrium there is no contribution from the free energies to the equilibrium phase-field profile, wheras for the case where different concentration fields are used for the respective phases, the concentration at a given point is written as an interpolation of the individual phase concentrations, and the equation is closed with the condition of equilibrium chemical potential among the phases, or alternatively a known partition relation among the phase concentrations which enables the determination of the phase concentrations. These are then utilized in the determination of the driving force for phase transformation. The common basis for both methodologies however, is that the driving force for phase transformation is the difference of the *grand potentials* of the phases, at the same chemical potential. Through this construction, it is evident that at equilibrium, there exists no terms arising from the chemical system, which contribute to the solution of the equilibrium phase-field profile, implying that the equilibrium properties such as the interfacial energies can be fixed independently of the free energy of the respective phases.

With this motivation a new model is derived, starting from a grand potential functional instead of the free energy functional (previous models) with the thermodynamic variable as the chemical potential instead of the concentration field. It is shown that, with this modification it is possible to get rid of the excess contribution to the interface and the length scale related to the interface thickness, is then independent of the chemical system one is simulating. This provides for significant flexibility in the applicability of the model for different alloy systems. This work bears co-incidental resemblance to the work by M.Plapp [91]

# 1.3. Asymptotics

In performing quantitative simulations, it is essential to acquire knowledge of the mapping of the model to the respective free boundary problem one is attempting to solve. To achieve this, we require to perform an asymptotic analysis. In this context, there are two limits, namely: the *sharp interface limit* (interface thicknesses tending to zero) and the *thin-interface limit* (interface thickness, remains finite, but small in comparison to the diffusion length). While the sharp-interface limit is relevant when the simulations are performed with very small interface thicknesses, the more technically relevant one is the thin-interface limit, since this allows one to retrieve the same free boundary problem but with length and time scales that are computationally accesible.

One of the key parameters, to fix in phase-field simulations, is the relaxation constant $\tau$ which relates to the relaxation of the interface. In most, mesoscopic simulations however, one is interested, only in the diffusion controlled growth regime. This implies, that the phase boundaries relax infinitely fast when imposed with a change in the coupled concentration field. This is achievable in the framework of a time dependent free boundary problem in the thin-interface limit previously derived by Karma [48]. The principal result states, if one derives, the expression for the interface kinetic coefficient, in the thin-interface limit, there exist parameters such that vanishing interface kinetics can be achieved. Whereas, in literature we derive that, this has been performed for the case of the double-well type potentials, in the present work, the thin interface limit is extended for the case of double-obstacle potentials.

While this limit makes effective use of thicker interfaces, there are some assosciated problems. With the use of thicker interfaces, there exist certain corrections one must include in the asymptotics to simulate the right free boundary problem [6]. These have been elaborately evaluated for both solutal and thermal problems, but for normally used double well potentials. The principal result in the analysis concludes that among the three thin-interface defects, two of them: *surface diffusion* and *interface stretching* are simultaneously absent, if odd-interpolation polynomials are used for interpolation of the diffusion constants of the phases and the free energies. However, such a choice makes it impossible to get rid of the third interface defect which is *solute trapping*. Solute trapping is a chemical potential jump at the interface resulting from asymmetric diffusivities of the two phases. Present models use a non-variational approach of using an anti-trapping current [46], to remove this jump at the interface. This has however been derived only for potentials of the smooth well type. In this analysis, the correpsonding expressions for the thin-interface kinetic coefficient and the expression for the anti-trapping current are derived for the the case of the double obstacle potential. Following is the summary of the goals achieved [18]:

- Development of a model based on grand potential functional

- Removal of additional limiting length scale resulting from variation of grand chemical potential excess at the interface

- Equilibrium properties such as surface tension are independent of the chemical free energy of the system

- Thin-interface asymptotics for the case the *double obstacle* potential

- Derivation of the kinetic coefficient in the thin-interface limit

- Derivation of the anti-trapping current for the double obstacle potential and a multi-component system, with vanishing diffusivity in the solid

## 1.4. Validation of the model

The model and its modifications are tested for real alloy systems. First, is the investigation of dendritic growth in the Al-Cu system. Comparisons are made with the analytical dendritic growth theories (LGK) and good aggreement is achieved. Among binary and ternary eutectics, the asymptotics and the model are tested with respect to theorectical expressions derived for coupled growth derived previously. In addition, the model is applied for the case of the Al-Cu-Ag alloy for the modeling of three-phase eutectic growth. A generalized route for the construction of free energy data, utilizing the essential information from databases, is constructed. Using this, some preliminary morphologies in 2D are presented at growth conditions relevant in directional solidification experiments.

# Chapter 2

# Phase field modeling of multicomponent systems

## 2.1. Introduction

In the past decades, the phase field method has become an important tool to describe microstructure evolution during phase transformations. In particular, considerable advancements have been made in the field of modeling phase evolution in multi-component systems. The areas of application though initially limited to solidification have spread to many a phenomena, involving solid-state diffusion, deformation behavior, heat treatment, re-crystallization, grain boundary pre-melting, grain coarsening etc. The phase-field approach's popularity is due to the elegance with which it treats moving boundary problems earlier in the regime of sharp interface methods. The interface representing the boundary between two mobile phases is replaced with a smoothly varying function called a *phase-field*, whose change represents phase evolution. This approach obviates the necessity to track the interface and hence makes large scale simulations of microstructure evolution involving complicated geometrical changes computationally tractable. The application of the phase-field method starts with the creation of the functional which includes the material properties involving both the surface properties of the interfaces in the system and the thermodynamic energy of the bulk phases in the system. A variational derivative of this functional with respect to any of the changing phase-field variables, gives us the driving force for the change. Depending on whether we are treating a pure component or multi-component system this driving force is a function of just the temperature or includes the compositions of the different components in the system also as variables. The source of thermodynamics of the bulk phases, are derived in a number of ways, starting from ad-hoc methods to creation of simpler thermodynamic models or the direct use of the well known *Calphad* databases.

This review is an attempt to gauge the applications of the phase field method in simulating the processing situations involving multi-component materials. Such a review would however be incomplete without a prior mention of the sequence of developments in the phase field method which has made such applications possible. In the following, we list some of these landmark developments and try to put them in context of the final goals that were achieved.

## 2.1.1. Model evolution

The phase field method originated as a branch of continuum theory and various instances have appeared in literature [14, 35, 39], while the first formulation of phase field equations to describe solidification can be found in the works by Langer [63] which were in turn based on the Model C of Halerpin, Hohenberg [39]. Similar independent models on pure metals were proposed by Collins and Levine [65], building up to the first large scale simulations of dendritic growth from pure melts which were performed by Kobayashi [61]. In these earlier works, the free energy formulations are ad-hoc, and motivation for the formulation of thermodynamically consistent models led to evolution equations being derived from a single Lyaponov type entropy functional, ensuring local maximization of entropy, [88, 124, 129] for two phase binary alloy systems. Later similar consistent models incorporating a generic formulation for treating multi-phase, multi-component models were proposed by Nestler and co-workers[33, 79]. However, the first multi-phase models are found in an earlier work by Steinbach et al. [108], where the free-energy formulation is ad-hoc.

Following the investigations of Caginalp and co-workers, [13] to relate the phase-field evolution equations to the sharp interface free-boundary problem along with the analysis of the attempts at quantitative comparison by Wang and Sekerka [123] and Wheeler et al., [136], makes it quite clear, that the parameters such as the kinetic coefficient and the capillary length derived from the sharp interface limit (the interface width going to zero) leads to quite stringent restrictions on the interface width for quantitative simulations, which in turn presents immense challenges on the computational side. Accompanying this problem, is the simulation of structures at the small interface kinetics limit, i.e., growth structures at low undercooling. In a time dependent free boundary problems (TDFBP), apparently it is evident that if one goes by the sharp interface limit, the kinetic coefficient will remain finite and hence, only simulations with interface kinetics can be performed. Solutions to this were extended by Karma [48] through the concept of *Thin interface limit* where the interface width remains non-zero, but much smaller compared to the mesoscopic diffusion length of the problem, which is the most relevant for the problems with a Stefan boundary condition. Although, the study is for pure metals where the phase evolution equations is coupled with the temperature field, later it is extended to the case of binary alloys, with constant partition coefficient and diffusivity through combined work with Losert and co-workers [72]. While this particular formulation is for a particular choice free energy density, a general procedure for the mapping of any phase-field model to its sharp interface limit is described in the investigation by Provatas et al. [28].

Dendritic solidification in the whole range of undercoolings still remains a challenge. It has been established that if the thin-interface correction is properly incorporated most phase-field models irrespective of their thermodynamic consistency, converge identically with similar computational costs for a given range of undercoolings. This range is identified as region where the *interface peclet* number (ratio of the interface width and the diffusion length) is small [59]. The convergence of a given phase-field model depends on the particular formulation, however, all phase-field models deviate from the intended sharp free boundary problem for higher *interface peclet* numbers. Since, at a given undercooling and strength of anisotropy, the velocity is fixed, the only degree of freedom that remains for adjusting the interface peclet number is the reduction of the interface width. This proves to be computationally expensive for large undercoolings (small diffusion length). For lower undercoolings larger interface widths can be used, but larger simulation domains are necessary which require efficient computational algorithms. To this end, *adaptive mesh* methodologies [94] and *random-walker* algorithms [92] prove quite helpful.

In the application of the phase field method, the WBM (Wheeler, Boettinger, McFadden) model [134] and the related formulation of free energy densities became quite popular. Kim [56] however, showed the existence of a potential excess, arising from the variation of the grand chemical potential across the interface at equilibrium, which is also previously illustrated in the works of WBM. The novelty is, that they show how this excess from the particular formulation of the free energy restricts the choice of the interface width for a given surface tension. Since the resolution of the interface is related to the domain that can be simulated (in a regular grid structure), this presents serious challenges to the computations that can be performed. Tiaden etal. [114] propose a solution to this problem by adopting different concentration fields for the solid and the liquid, connected to each other through a partition coefficient, which is a function of velocity. This is also thermodynamically consistent for dilute binary alloys. Kim and co-workers [56] provide an extension by also considering the concentration at a point to be a mixture of concentrations of the solid and liquid phases, but the individual compositions are corresponding to the parallel tangent construction between the free energy curves of the respective phases in contact. Both methods relax the restriction placed on the choice of interfaced widths. Models following similar lines as Kim, have also been proposed by Cha et al. [16]. Later, some more related variational and non-variational methods of getting around this problem were formulated by Plapp et al. [31, 32] for two phase eutectic solidification, without the use of separate concentration fields for the solid and the liquid.

The second problem is related to the choice of thicker interfaces which leads to the modification of the Stefan condition. The aggregate of modifications to the Stefan boundary condition due to the choice of a finite interface thickness came to be called *Thin Interface defects* and a rigorous mathematical description of each of these defects are found in investigations by Almgren and Mcfadden, [6, 75] for pure melts with asymmetric diffusivities for the transport of heat, while similar effects are also shown to exist in alloys [46]. The thin interface defects discussed in these studies are three in number. Of them, the defect of solute trapping was earlier discussed in the work by Kim and Ahmad, [1, 55]. In real materials the diffusivity of the solid is much lower compared to that in the liquid, which is different from the assumptions by Losert et al. [72]. The asymmetry in diffusion constants causes the asymptotic limits of the chemical potentials at the interface, resulting on the liquid and solid side of the interface, to be different. This difference is a function of velocity and the width of the interface, causing a chemical potential jump at the interface also called *solute trapping.* Although this is a phenomena seen to occur in materials science during rapid solidification, the difficulty arises from the fact that the magnitude of the solute trapping effect scales with the interface widths, which being chosen orders of magnitude higher for the phase field simulations, than that in a real material, gives rise to significant solute trapping at velocities where it would be negligible in a real alloy. The second, known as the *surface diffusion* resulting from the mismatch of the fluxes on the solid and liquid sides of the interface, and the third is *interface stretching* that results because for a solid growing with a convex interface into the liquid, the source of the solute on the solid side is over a smaller area compared to that on the liquid side because of the curvature of the interface, over a finite thickness. The physical interpretations of the defects are elaborated in the explanations on quantitative modeling by [30]. It is illustrated, that the solution to the problems, is related to the appropriate choice of interpolation polynomials. However, due to the restriction of the number of interpolation polynomials, it is generally agreed that all three effects cannot be simultaneously taken care of, while maintaining reasonable bounds of computational efficiency. Presently there exists no phase-field model which removes all three effects using a variational formulation. Therefore, while the problems of surface diffusion and interface stretching can be corrected through the use of suitable interpolation polynomials, the problem of solute trapping is removed using a non-variational formulation as suggested by Karma et al. [30, 46], by using an *anti-trapping current* which is a current of solute from the solid to the liquid. Another solution, to limit the thin interface defects, is proposed by Kim et al. [57], where they decouple the solute and the phase-fields, since the problem of solute trapping arises because of the variation of the solute field over a larger diffuse area than the interface. The diffusion field for the solute is limited to suppress the thin interface effects. However, in practice it is found that in calculations, one needs to use a diffuse length of the solute field slightly larger than the grid size $\Delta x$, which makes the solute trapping effect still appreciable. Later, the earlier phase-field model of Kim et al. [56] was extended to multi-component systems,[54], with the additional removal of the thin interface defects using an anti-trapping current formulation. Provatas et al. [84] present an extension of the model of Eschebaria et al. [30] to treat multi-phase binary alloys.

The complete problem of microstructure evolution in material science normally involves the coupling of the thermal and solute fields. This coupling is however a challenge in simulations, as the heat and solute diffusion operate on different time scales with the thermal diffusion being much faster. Phase field modelling of this coupled phenomena was first performed by Boettinger and Warren [11], where they ignore the spatial variation of the temperature field, and the temperature field is computed from heat balance equation between the imposed heat extraction rate and the released latent heat. It was found later that this is valid only for low undercooling [71] where the authors compare the temperature fields computed by exactly solving the diffusion equations for the internal energy, and those from average heat balance equation coupled with the isothermal model. The model proposed however, has thin interface effects, which are absent in a later model proposed by Ramirez and Beckermann [99]. The diffusive time scales for heat and mass transfer are comparable during rapid solidification of alloys and hence the problem becomes more tractable. The thermal and solute-field coupling in rapid solidification problems are treated in the works by Conti et al. [21, 22].

Along with the massive developments to make quantitative simulations of real materials feasible, model adaptations to treat other phenomena, like nucleation were proposed by Granazy et al. [37, 115, 131] in solid-liquid transitions, while in solid state precipitation, similar models were formulated by Simmons and co-workers [106].

It has long been the aim of the phase field community to treat real materials and the significant step towards this, comes through the use of free energy of the different phases, directly from the proven and tested CALPHAD databases. The first attempts are found in the model adaptations proposed by [15, 36, 96]. In recent works, the usage of the CALPHAD databases, and its direct coupling to phase-field solvers has become more frequent and has spread to a variety of applications, which will be highlighted as we take an overview in the later sections.

In the following sections, we present the various applications of the phase field models. In the first section, we list the studies using idealized model systems constructed for studying the physics of a particular solidification process, followed by applications where real thermodynamic databases were used for the simulations in various processes and concluding with an outlook for the application of the phase field method to different phenomena and resolving of other research issues.

## 2.2. Phase-field applied to problem of solidification

Solidification is the phenomena most extensively studied using the phase-field method. The types of solidification reactions range from pure metal solidification, eutectic (liquid on solidifying giving rise to one or more solids), peritectic (liquid on reaction with a solid gives another solid) and monotectic solidification (liquid on solidification gives a solid and another liquid). Each of these types have been explored by the phase-field method. In this section, we list down the studies where the central point is to understand the physics accompanying the evolution process. Most investigations in this section involve the use of free energy densities which are created in an ad-hoc manner or are generated through simpler thermodynamic models which do not influence the final inferences.

## 2.2.1. Dendritic solidification

Dendritic solidification is ubiquitous in materials science and has long intrigued materials scientists and physicists, as to what are the parameters and conditions, which lead to this instability. Although a lot of understanding has been gained about the physics of this effect, simulations or experiments are necessary to characterize the materials response to processing conditions. In this regard the phase-field simulations come to be of much use.

Beginning with the first large scale simulation of thermal snow flake dendrites by Kobayashi [61], parallel attempts at quantitative comparisons were made by Wheeler et al. [136], where the morphology of dendrite tip is compared to needle crystal solutions proposed by Ivantsov for pure Ni dendrites. Along with this, investigation of the operating state of the dendrite tip and matching with the marginal stability criterion and the micro-solvabilty theories is also carried out. The first simulations of solutal dendrites in a Ni-Cu system are found in the studies of Warren and Boettinger [130] who employ a thermodynamic consistent model for the investigation. Fig. 2.1 shows an illustrative phase-field simulation of a three dimensional (3D) Al-Cu dendrite. The thin interface limit for crystallization of pure materials [48, 50] earlier proposed by Karma for pure metals and later extended along with Losert and co-workers [72] for a binary alloy (Succinonitrile-Coumarin), is used for studying the range of wavelengths for the formation of stable singlets, doublets, and unstable transient patterns on perturbation of a planar interface. These works illustrate that one can choose an interface width in the mesoscale range and still perform phase-field simulations independent of interface kinetics.

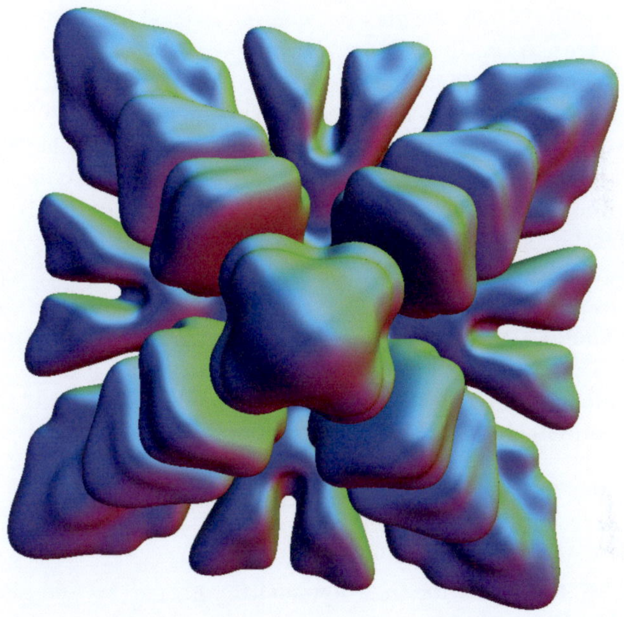

**Figure 2.1.:** AlCu dendrite simulated using a thermodynamically consistent phase field method, with fourfold cubic anisotropy. Secondary arms are initiated through noise. An ideal solution model was assumed for the bulk free energies.

Thus, quantitative simulations become theoretically possible also in the range of low undercooling. These studies demonstrated the usefulness of the method, however, quantitative applications to real materials are only possible with the introduction of the anti-trapping current [46] along with solution to the other thin interface defects that arise due to asymmetric diffusivities in the phases [6]. Parallel to these developments, the phase field method has been applied to different areas of materials research, such as solidification of bulk metallic glasses, where it is used for studying dendritic to globular transitions in ternary $Ni_{60}Cu_{40-x}Cr_x$ [25]. A morphology map, showing the change in the microstructure from dendritic to globular morphology, as a function of the Cr concentration is derived. Concentration profiles, measured in experiments and computed in phase field simulations during solidification of multi-component metallic glass composites, are compared and good agreement is found in the works by Huang et al. [43] and Nestler et al. [78]. The phase-field method is also used in the simulation of dendritic solidification in industrial alloys, where higher order Redlich-Kister polynomials are used by Wang [121] for the description of the free energies of the hexagonal $\alpha - Mg$ phase and the melts in the AZ91D Mg rich Al-Mg alloy. Both thermal and solute fields are solved and a 3D hexagonal anisotropy in the kinetic coefficient and the surface energy, reproducing stacking of 2D hexagonal plates is proposed.

## 2.2.2. Eutectic solidification

Eutectic alloys are useful for their low melting properties and also for their mechanical properties given the structure is uniform at the finest scale. Hence, a study of these alloys to gain an understanding of the relation between the processing conditions and the final microstructure is useful for material scientists. It is also an interesting topic for physicists because of the number of possible pattern formations. The first phase field model, enabling the treatment of the transformation of a liquid to two solid phases was proposed by Karma, [45]. The model uses the concentration field as an order parameter to distinguish between the solids, similar to the works of Cahn [14], while a second order parameter is used, in order to distinguish the solid and the liquid phases. The free energy surface is created in such a manner that the solid free energy contains two minima symmetrically placed with respect to the eutectic composition and corresponding to the two solid phases in the system. A directional solidification set up is created, by setting a temperature gradient in the growth direction and the equations are solved in the moving box frame, which is set at an imposed velocity. Comparisons of the angle at the triple point are found to be in good agreement with the predictions derived out of the Young's condition in the sharp free boundary problem. The results of the average interface undercooling and the minimum undercooling spacing match well to the predictions of the classical *Jackson-Hunt Theory*. Similar models, with extensions allowing the free energy density of the solid phases to differ in their form are proposed by Wheeler et al. [135].

Athough stable coupled growth is commonly observed during eutectic solidification, they also exhibit certain instabilities during growth of thin and bulk samples. In this context, phase-field models provide a great means to study the instabilities in the growth patterns of eutectics which is a big development over the *boundary integral method*, which is unable to treat catastrophic changes, like lamella elimination or termination. Stability of 2D eutectic patterns to lamellae elimination is investigated by Akamatsu and co-workers [4, 5]. Measurements of the average undercooling of the growth interface from directional solidification experiments of eutectic Carbontetrabromide-Carbonhexachloride organic alloy in thin film morphology agree well to calculations from simulations. The studies reveal that configurations of lamellae with average spacing below the minimum undercooling spacing can also be observed both in experiments and simulations. This is however in contradiction to the result which is obtained, if we combine the analysis of Langer [62] with Cahn's earlier hypothesis, which states that the growth of the lamellae is always normal to the local solidification front. Langer uses Cahn's hypothesis in his stability analysis of lamellar growth, for small amplitude long wavelength perturbations in the spacing of large arrays of lamellae. The outcome of the analysis is that configurations with spacings smaller than the minimum undercooling spacing are unstable and will eventually die out. Akamatsu and co-workers reason this anomaly in their work by relaxing Cahn's growth condition of the local growth velocity being normal to the interface and put forth a stability analysis, which provides reasoning for the observation.

The presence of a third impurity component in a binary alloy is known to destabilize a planar two phase eutectic front giving rise to colonies. Plapp and Karma [93], study this effect of colony formation using the phase field method and obtain qualititative comparison with their previous stability studies. In the simulations, they show the breakdown of a solidification front comprising of two solids into two phase cells. There was however, no steady state envelope found and the structure exhibited tip splitting and cell elimination events until the very end. It is worthwhile to note that even in these studies of colonies, Cahn's hypothesis of growth remaining normal to the solidification front is weakly violated and the authors believe that this has effects on the stability properties of the eutectic front.

The lamellar eutectic growth regime has been shown to exist in a finite range of spacings around the minimum undercooling spacing. Beyond a threshold spacing, the patterns bifurcate to different oscillatory patterns 1-$\lambda$-O and 2-$\lambda$-O along with tilted states, and also with mixed instability modes. While the pure oscillatory modes and the tilted states were studied by Karma and Sarkissan using a boundary integral method [51] experimental observations of mixed modes in the presence of capillary anisotropy are found in the works of Ginibre et al. [34]. The first simulations with a multi-phase, multi-component field model by Nestler et al. [81], illustrate different oscillatory patterns in 2D, while illustrations of the 1-$\lambda$-O and 2-$\lambda$-O modes are shown in Fig. 2.2. The N order parameter approach is an elegant one

 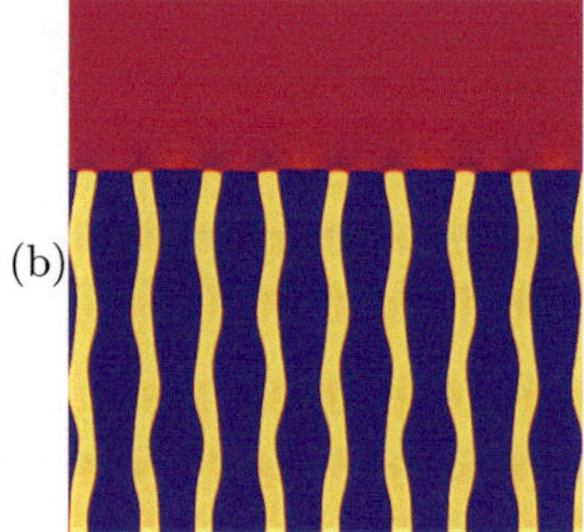

**Figure 2.2.:** Illustrative 1-$\lambda$-O (a), and 2-$\lambda$-O modes seen in binary eutectics.

for treating such multi-phase problems. A systematic investigation of these instabilities using the phase field method is performed by Kim et al.[58] and a morphology map is presented based on the composition and the spacing, while continuing works in 3D showing the possibility of a zig-zag instability are found in investigation by Plapp and co-workers [87, 90]. These morphologies are also seen in experiments [2]. More recently structures showing shape morphology transition from rods to lamellar structures, depending on the composition of the liquid are found in simulations of 3D directional solidification performed by Parisi et al. [86]. In directional solidification of a eutectic alloy the microstructure that the system chooses is a function of the composition in the liquid, among other things. In 3D a number of structures can be found ranging from rods, to lamellae, in-between structures, with a mixture of two, and also defect structures such as elongated rods are possible.

## 2.2.3. Peritectic solidification

Peritectic solidification is an important topic to understand because of the important materials in industry derived through this process. Most of the materials are for magnetic and superconducting applications, while a number of them also are useful for high strength applications, for instance certain super alloys like $Al_3Ni$. A number of microstructures have been found to exist during peritectic solidification. Among them, include the *engulfing* microstructure, in which the peritectic phase grows over the pro-peritectic phase in the form of spherical nuclei, or infinite plate. These structures have been first studied by Tiaden et al. [114] in a Fe-C alloy and then later by Nestler and Wheeler, [81]. In these studies, the nucleation events of the peritectic phase are explicitly put, and the question of nucleation sites is unclear. In a recent study [19], an attempt has been made to resolve this question. Along with this a prediction has also been made on the lower bound of the solid-solid surface tension for which the *engulfing* microstructure is found to exist.

Researchers have long been trying to understand the phenomena of the formation of different microstructures such as the *island growth, banded structures*, and *coupled growth*, and although theories have been proposed on the mechanisms leading to the formation of these structures, no quantitative predictions have been possible. The stability of these morphologies, namely the island and banded structures is examined using the phase-field model in the work of Lo et al. [69]. The study reveals that the formation of island and banded structures occurs in the hypo-peritectic region in the absence of convection, which also corresponds well to existing theories. Below a concentration limit of the solute and when the lateral size is below a critical value, island structures were found to be stable else banded structures in 2D are found to evolve.

Coupled growth is another interesting phenomena found in experiments, which has been studied using a phase field model [26]. The authors find that coupled growth occurs for select range of parameters and they present a morphology map showing the regime of coupled growth as a function of the concentration of the liquid, and the G/V ratio (Thermal Gradient G, Velocity V). In isothermal solidification coupled growth is always preceded by the formation of islands giving way to 1-$\lambda$-O oscillations, followed by quasi-steady coupled growth. In non-isothermal situations (with a temperature gradient), for a given concentration in the liquid, a range of spacings $\lambda_{min} < \lambda < \lambda_{max}$ exist, between which coupled growth is achieved. Outside this range, the structure exhibits oscillatory instabilities. The striking observation is that, coupled growth structures are also found in regimes where $\dfrac{\partial T}{\partial \lambda}$ is less than zero (Where T is the undercooling). According to Cahn-Jackson-Hunt theories, coupled growth will be unstable in the region where the undercooling reduces with increase in lamella spacing. However, as is also found in the case of eutectic coupled growth, the gradients in spacing can also be relaxed through lateral motion of the tri-junction points in the presence of a thermal gradient and the growth envelope is not strictly normal to the local interface shape. The authors believe that the same mechanism is also qualitatively responsible for the observation of coupled growth in peritectic solidification i.e. the presence of a temperature gradient provides a stabilizing force which counteracts the force, causing the engulfing of a lamella.


## 2.2.4. Monotectic solidification

Monotectic alloys are characterized by the monotectic reaction at a fixed temperature where a liquid on solidifying gives another liquid and a solid. The phase diagrams of these systems, have the property that towards one side of the invariant point, (hyper monotectic region), there exists a miscibility gap in the liquid. Phase separating liquids are detrimental and hence production technologies are normally designed such as to avoid this process. The alloys are however useful in applications such as self-lubrication. To control the phase separation and the solidification microstructure, one needs to have a proper understanding of the evolution kinetics and the processing parameters affecting a given morphology. Since the process of monotectic solidification involves the interaction of two fluids of differing concentration, a number of physical phenomena resulting from the coupling of the fluid properties with the diffusion and capillary effects influence the final microstructure. The first phase field method used for studying this reaction is by Nestler et al. [82]. Two models are proposed, differing in the freedom to choose the temperature ranges that can be treated. One of the models, treats the free energy of the liquid such that it is non-convex for a given concentration range, hence requires an energy term proportional to the gradient in the concentration fields in the functional, to stabilize evolution when concentration of the liquid is inside the spinodal decomposition regime (phase-separation region). This is computationally more expensive because the concentration evolution requires the solution of a *bi-harmonic operator* in contrast to the laplacian in case of linear diffusion. The second model decomposes a single free energy density for the liquid, into two parts, each of which is convex. This formulation is more suited for the problem of solidification, while it is not suitable for treating the phenomena of phase decomposition.

The model also includes terms to treat the *marangoni* effect. This is formulated by treating the complete capillarity tensor for application in the flow-field evolution equations. This includes the effect of the concentration gradients on the surface energy at the interface between fluids, and the resulting flow, due to a gradient in this term. Good qualitative comparisons are made with experiments in the directional solidification setup.

Later works by Tegze et al. [112] investigate all the diffusion and hydro-dynamic effects affecting the droplet distribution sizes in the Al-Bi alloy, using a regular solution model for describing the free energy of the phase separating liquid. Figure 2.3 shows exemplary phase seperation of pure liquid into two separate liquids at two different compositions.

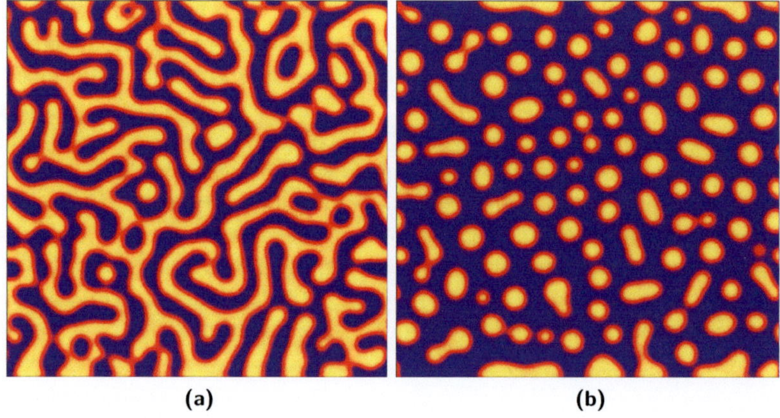

(a)        (b)

**Figure 2.3.:** Phase seperation at two different compositions of the liquid in a model Fe-Sn alloy. In (a) the simulation is performed at the monotectic composition while in (b) a hyper-monotectic composition is employed.(Work along with co-worker Wang Fei)

## 2.2.5. Nucleation

The understanding of the phenomena of nucleation is critical to having a control on the length scale of microstructure. The process of nucleation being a stochastic event can be treated in the deterministic framework of the phase-field method, through two methods: incorporation of noise, mimicking the thermal and concentration fluctuations in the system or explicitly depending on a criterion of nucleation derived from the functional. The phase-field method provides an excellent method for deriving the properties of the nucleus. The *nucleus* in this framework, refers to the phase-field profile which satisfies the *Euler-Lagrange Equation* under the constraint of constant chemical potential [37]. The resulting phase profile is an extremum of the free energy functional. The grand chemical potential excess with respect to the initial liquid calculated using this profile, is the *barrier to nucleation*. Such a theory has been postulated for both homogeneous and heterogeneous nucleation [131]. These calculations have been verified also with atomistic calculations. Additionally, variation of quantities such as the *Tolman length* (difference between the radius of the equi-molar surface and the radius of the surface tension) predicted by the atomistic calculations, have been qualitatively reproduced in the simulations [109, 115]. Large scale simulations with concurrent nucleation events have also been performed for systems like Ni-Cu [95] and technically relevant alloys such as the Al-Ti, where transition from Columnar to Equi-axed morphologies (CET) is simulated. An exemplary structure obtained by incorporating noise in the simulation domain is shown in Figure 2.4.

## 2.2.6. Solid-State

In this subsection we look at phenomena occurring during solid-state transformations, investigated using the phase-field method. In the survey, we are going to concentrate especially on a topic which has been of extensive research interest over the last 15 years namely the processing of Ni-based super alloys.

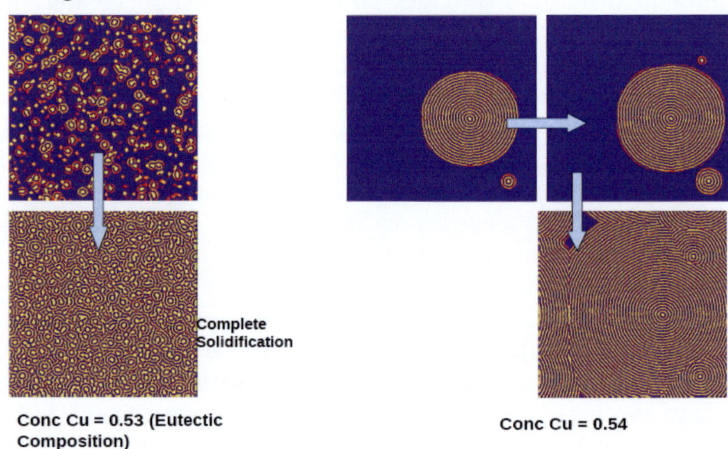



**Figure 2.4.:** Nucleation structures obtained in a model Al-Cu alloy at two different composition in the liquid. While a uniformly chaotic structure is obtained at the eutectic composition in (a), a more regular structure is obtained at a off-eutectic composition in (b).

## Super Alloys

The term *superalloy* was first used to refer to the group of alloys for use in turbine superchargers and aircraft turbine engines that require high performance at elevated temperatures. Presently, the application of such alloys has extended also to land-based gas turbines, rocket engines chemical and petroleum plants. A characteristic feature of these applications, is that the energy efficiency goes up with temperature. Understandably, the need arises to maximize the temperature of application. Typically, the intended temperature of use is up to 90% of the melting temperatures. At such high temperatures and applied stresses, failure mechanisms such as *creep*, hot corrosion etc. operate. This implies, that a material suited for such application needs to be resistant to such mechanisms of failure for extended exposure to such conditions. It has been found that alloys of elements in the Group VIII of the periodic table namely Fe, Ni, Cr, Co are particularly suited for this purpose. The alloys also involve smaller additions of carbide formers such as the W, Mo, Ta, Ti, and Al. In all, we can classify the set of super alloys into three classes namely, Ni-, Fe- and Co- based alloys.

One of the principal mechanisms of creep is *grain boundary sliding*. Thus, it is desirable to have them processed as single crystals (i.e. with no grain boundaries). With the invention of the efficient directional solidification techniques it is now possible to process them as single crystals, with the desired grain selectors. The quality of the alloy however, is related to the processing conditions and the purity with respect to undesired elements in the alloy. So one direction of research is to increase the purity of these alloys resulting as a process of solidification.

Research is also underway to get an understanding of the mechanisms of creep and deformation in such alloys to better predict the performance of the microstructure at elevated temperatures and stresses. This would help in engineering materials with required hardness and strength at these temperatures. There are two hardening mechanisms that are pertinent in such alloys, one is *solid-solution strengthening*, achieved with alloying additions and the other *precipitation hardening*. Solid-solution strengthening deals with the toughness, induced by the stresses due to the additional alloying elements because of the size differences between the solute and the solvent atoms. These internal stresses interact with the dislocations and act as barrier to their movement. However, this is relevant for low temperature ductility and toughness. The strengthening at higher temperatures occurs through precipitation hardening. The precipitates interact with the dislocations and pin them, preventing cross slip. One of the ways of strengthening in polycrystalline super alloys is the use of carbide forming elements which form carbides of the same crystal structure as the matrix and segregate to the grain boundary and pin them. The other type of precipitates are coherent and close chemical compatibility with the matrix. A unique phenomena associated with these precipitates is the increase of yield stress with temperature also called the *yield stress anomaly*. This is related to the locking of dislocations in *Anti Phase Boundaries* (APB's) (interfaces between possible ordered variants of different orientations), which lie on non-slip planes in most super alloys. This is a motivation to study the formation of such boundaries and their interaction with dislocations.

The strength of the alloy is related to the size of the precipitates hence research is also focussed, on getting an understanding of the coarsening kinetics of the microstructure as a function of the temperature, the alloying additions and the processing conditions such as the ageing temperature.

Given the number of experiments required and the high cost for the processing of these alloys, an optimization of the processing conditions through an extensive experimental study looks daunting. However, with simulations the task looks achievable, and promising. In this respect, the phase-field method through its evolution over the years has become a potent tool for the treatment of this problem. In the past decade, an alloy which has been particularly treated is the Ni-Al based super alloys, which are extensively used for turbine blades. In addition, it also suits the framework of the phase-field method, since thermodynamic databases of this alloy are available from CALPHAD along with information for mobilities from DICTRA databases.

In Nickel based superalloys, the microstructure is made of a disordered fcc matrix called the $\gamma$ phase and an ordered $\gamma'$ precipitate phase. The $\gamma'$ phase has the $L1_2$ cubic structure which is coherent and of high chemical compatibility with the matrix. The $\gamma'$ phase is the principal phase responsible for precipitation hardening, and hence has received a lot of attention over the years. The $L1_2$ variants are four in number and hence in a binary

Ni-Al system, the system can be defined using a concentration field and three long range order parameters. In earlier works by Khachaturyan et al. [126] the formation of squared precipitates is already achieved in simulations, however a major drawback is that the matrix phase and the precipitates are treated iso-structurally i.e. the four variants do not have an energy barrier between them. While this is true for certain precipitates for instance $\delta'$ precipitate in Al-Li alloys and the matrix, this is not the case of for the $\gamma'$ precipitate. As a result of the simulations performed by various authors [8, 103, 125], the conclusion is that the treatment of this energy of the anti phase boundaries is necessary for retrieving the anomalous decrease in coarsening rate at higher volume fractions which is also seen in experiments. The authors define two terms *in-phase* and *out-of-phase*, implying the case when two variants of the same type are next to each other and the other case when they are not respectively. The *out-of-phase* particles on coalescence form Anti-Phase Boundaries (APB's) which are for the case of Ni- alloys, twice the energy of the interface between a variant with the matrix. This being the case, two *out-of-phase* particles will never coalesce forming an APB when both are in equilibrium with the matrix.

There are two coarsening mechanisms in such alloys, one is through a diffusive process called the Ostwald ripening, where the smaller particles reject atoms and shrink, while the larger particles absorb and grow. While this mechanism of interface energy minimization is present in all alloys it is of a longer time scale, than a second mechanism which occurs through displacive-coalescence mechanism. The displacive-coalescence mechanism however, occurs only if two precipitates next to each other are in-phase. With increase in number of particles, the probability of two particles being in-phase is very small and hence most of them are out-of-phase. Hence, with increasing volume fraction of particles the coarsening rate is reduced. Additionally, *split* microstructures seen in experiments can also be explained with this mechanism, where the coalescence of in-phase particles to minimize energy is the driving force for the formation of such patterns [8]. Incidentally, there appears to be no reason to believe that the coalescence should occur above some critical size of precipitates as previously believed, as such a mechanism of coalescence of in-phase particles is able to explain all observed experimental microstructures. A detailed study of the coarsening kinetics in the different regimes of volume fractions and particle sizes is present in the work by Vaithyanathan and Chen [117]. Under conditions of stress and high temperature, the precipitates align themselves parallel to the stress direction in the case of tensile loading and normal to the direction of tensile loading, for the case of compressive loads. This phenomena is called *rafting* or alternatively, *directional coarsening*. In experiments, the interrupted stripes of precipitates are observed and here again it has been verified in simulations [125, 127] that incorporation of APB interface energies in simulations is necessary to observe discontinuous aligned set of precipitates, instead of continuous stripes. The former is also seen in experiments.

It is clear that the degrees of freedom of such a system, are the interfacial energies, the bulk free energies and the lattice mismatch. In the last decade the emphasis has been to give more quantitative meaning to simulation results. In this direction the linking of the simulations to CALPHAD databases has been useful. The databases provide free energies using the sub-lattice models. A four sublattice model is used for the case of treating the variants. The site fraction in the CALPHAD databases can be used interchangeably as the order parameter in the phase-field setup. This gives us a continuous free energy in the order parameter/site fraction and the composition space. The first attempts of linking the databases are able to retrieve, the time evolution of the precipitate morphologies, from spherical at small sizes which occurs as a result of minimization of the surface energy, while as the precipitate becomes large the particles change the shape to cubical which occurs via minimization of the bulk elastic energies [141]. The elastic misfit strain energies are in all these studies treated as some derivative of the *Vegard's* law. The only degree of freedom remaining, is then the interfacial energies. While this quantity is difficult to get experimentally, models such as the Cluster Variation Models (CVM) have been used to get an idea of the composition profiles, phase boundaries and interface energies [119]. The results of such calculations have been able to give predictions which agree well with experiments. With the growth of CALPHAD and DICTRA databases in recent years, simulations of new generation Ni- based super alloys is becoming possible. Concentration profiles, occurring during heat treatment and temperature processing in multi-component Ni- base super alloys are calculated by some groups [60, 120, 132, 140], giving good agreement with experimental measurements. For the case of non-isothermal treatments such as heat treatment one would need to incorporate new precipitated nuclei as the temperature is reduced. To achieve this, Langevin noise is used by Simmons et al. [107] and an interesting analysis of the process of aging during heat treatment is also performed by Wen et al. [133]. Exemplary structures during aging and ripening are simulated for the alloy Fe-Cu and are shown in Figure 2.5. While in our discussion we have stated applications of the

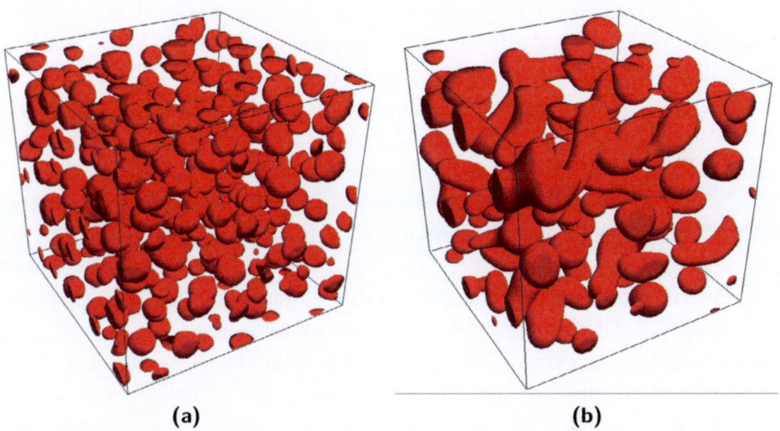

(a)          (b)

**Figure 2.5.:** Aging in a Fe-Cu alloy. In (a) is the starting condition, while (b) shows the state after some particles have co-agulated and some have ripened.(Work along with co-worker (Rajdip Mukherjee))

phase-field method only for Ni-based alloys, some similar studies have also been performed for the case of other super alloys such Ti-Al-V [17].

The phenomenon of precipitation is not only interesting in super alloys, but some common alloys like Fe-C. However, precipitation is not always useful like in the case of *Widmanstätten ferrite* which grow as needles. Knowledge of the conditions which result in these microstructures and their control, are important for steel alloy design. Phase-field modelling is applied for studying such structures [70] using CALPHAD databases for the free energy of the bulk phases.

## 2.2.7. Other fields of application

Metal forming is an important process in any processing chain. A control of the grain structure during high temperature forming decides the resultant strength of the material. During forming at elevated temperatures one of the most important processes which occur is *re-crystallization*. Most of the last decade this process has been modelled using the combination of the *crystal plasticity method* for treating the stresses and strains during deformation and re-crystallization is solved using the *Cellular Automaton Method* [97, 98] which describes specific rules of transformation for a grain, depending on the stored energy of a grain and the surrounding grains. While a number of theories have been put forth for describing the initialization of re-crystallization namely the misorientation or the energy stored in the grains, the actual criterion is still unclear. The process of growth after nucleation can however be well accounted for. Grain growth being a capillary force driven problem, is better suited for treatment by diffuse-interface methods such as the phase-field method. The application of this method is however still in its in early stages with first attempts by Takaki and co-workers [110, 111]. They treat both static and dynamic re-crystallization. In the treatment of static re-crystallization [111], the stresses are computed from the crystal plasticity method, which are used to determine the sub-grain structure. The diameter of the sub-grain structure is calculated from the balance of the stored energy and that required for the creation of a new grain boundary. In the case of dynamic re-crystallization [110] dislocation density varies with time and the authors describe the evolution of the dislocation density as a function of the strain. The driving force is a function of stored energy in the form of dislocations and the criterion for nucleation of a new re-crystallized grain is derived using a *bubble model* described in the paper, which gives the critical density of dislocations required for the nucleation of the new grain. M.Wang and co-workers [122] simulate re-crystallization in a magnesium alloy AZ31 using the Gibbs free energy from CALPHAD databases. The authors solve the coupled problem of diffusion and re-crystallization.

Beyond the main streams of developments in solidification and solid-state transformations, the linking of phase-field models with thermodynamic and kinetic databases emerged as a powerful hybrid method also in other fields of applications. We briefly report on research work in liquid-liquid phase systems and grain boundary premelting.

A phase-field simulation study in [12] is devoted to investigate the dynamics and morphologies of spinodal decomposition of two immiscible liquids. The alloy of consideration is the binary system Bi-Zn, which contains a typical miscibility gap in the phase diagram. Different morphologies of the phase separation process are found at different regions of the phase diagram i.e. at different temperatures and concentrations. Furthermore, the authors describe a strong dependence of the microstructure on the formulation of the Gibbs energy, by comparing two different formulations determined by the CALPHAD method at the same temperature and concentration condition. A problem seen during continuous casting, namely hot cracking, is known to originate at the grain boundaries because of liquid films at grain boundaries. This phenomenon is called *grain boundary pre-melting*. The wetting occurs below the solidification temperature and cracks as a result of these pre-melted areas. A control of the temperature and parameters affecting the process is quite useful for avoiding material wastage. The classification of grain boundary pre-melting transitions in Cu-Ag solid solutions has been the focus of phase-field simulations in [76]. A multiphase-field model has been composed, with three phase-field parameters to distinguish two grain states in the presence of a liquid phase. Depending on the grain boundary energy, the temperature and the grain composition, a variety of pre-melting evolutions has been observed including (i) dry grain boundaries, (ii) completely wetted grain boundaries with pre-melted layers of diverging thickness, (iii) grain boundaries with discontinuities of the pre-melted layer thickness and (iv) metastable grain boundary states above the solidus line indicating the possibility of superheating (respectively supersaturation). The pre-melting behaviour is related to the disjoining potential combined with thermodynamic properties of the bulk phases.

# 2.3. Outlook

Despite the tremendous effort and rapid development in a broad range of multicomponent modelling applications, there still remains a comprehensive demand for further understanding of the physical mechanisms behind structure formation, involving the intensive investigation of the influences of external fields and processing conditions on the mechanical properties of materials used in technical applications. The existence of *just* a ternary component, e.g. in eutectic systems gives rise to a large diversity of new morphologies and illustrates the complexity amongst the regularity of nature. The computational study, particularly in 3D, holds a great potential for new insights into the physics of these materials. In combination with phase changes and solute diffusion of multiple components, another important challenge for future research will be the investigation of the effect of coupled fields such as fluid flow, stress and strain and plastic deformation. In all multicomponent modelling applications, the configuration of data sets and processing conditions will play a key role, essentially requiring the advancement of multi-scale and hybrid modelling techniques. The large amount of field variables to be solved as well as the desired large-scale 3D simulation domains will increasingly ask for employing modern computing methods including intelligent algorithms, parallelization on high performance computing architectures as well as optimized numerical solution approaches such as multi-grid and homogenization.

# Chapter 3

# Growth morphologies in peritectic solidification of FeC

# 3.1. Introduction

The peritectic reaction $L$(liquid) $+ \delta$ (ferrite) $\rightarrow \gamma$ (austenite) in Fe-C system occurs during the solidification of low carbon steels. A peritectic transformation (solid-solid transformation) and peritectic reaction is also observed in many other systems like Fe-Ni, Cu-Sn, Ni-Al, Ti-Al. The peritectic reaction in Fe-C is characterized by an appearance of the peritectic $\gamma$-phase that separates the liquid $L$ and the properitectic $\delta$-phase, followed by growth of $\gamma$-phase due to $L \rightarrow \gamma$ and $\delta \rightarrow \gamma$ transformations. Shibata et al. [105] experimentally observed the peritectic microstructure formations in the Fe-C system. They also formulated a mechanism for the peritectic reaction and transformation, based on the analysis of the kinetics of the phase changes. Various growth models have been proposed by others, to explain the peritectic reaction and transformation in different solidification morphologies [89].

In the past two decades, phase-field simulations have become a powerful tool to describe growth morphologies during complex phase transformations. The methodology has been used to model eutectic, peritectic and monotectic reactions, [81, 82]. In particular, the peritectic solidification of Fe-C was simulated using a multi phase-field approach [113], where the phase field equations were derived from a free energy functional and carbon diffusion equation was formulated on the basis of a separate solute diffusion model. A multi phase-field model was also used to numerically simulate the peritectic reaction by Lee et al. [64]. In previous work [81], a phase-field approach was formulated for both, binary eutectic and peritectic alloy systems. The model incorporated the free energies corresponding to a specific type or region of a phase diagram, which can be described through an ideal solution formulation. Due to the great similarity with respect to the free energies of the phases and accordingly with respect to the construction of the phase diagram, a unique formulation of a phase-field model was derived, capturing the solidification process in eutectic and peritectic systems, by setting up suitable values for the latent heats and melting temperatures. The approach was successfully applied to computations of various eutectic and peritectic growth structures, related to model alloy systems. Directional solidification of peritectic alloys, without morphological instability was studied using the phase-field method by Plapp et al.

[69]. The investigation also involved a study on nucleation and its effect on pattern formation. Phenomena of coupled growth and banded growth structures were also discussed in [47, 68]. Numerical studies on heterogeneous nucleation of the peritectic phase in the ternary system Nd-Fe-B were performed by Emmerich et al. [29] to examine the morphological effects on nucleation in 2D.

In this chapter, we investigate different 2D and 3D growth morphologies during peritectic growth in the Fe-C system using a multi-phase, multi-component phase-field model. We also investigate the effect of surface energies and evolving concentration domains on nucleation and growth behavior. In the next section, we give all the mathematical functions and dynamical equations of the phase-field model and explain how the parameters are calibrated. Section 3 is devoted to the discussion of three broad classes of growth morphologies of the peritectic reaction, containing a study of the evolution characteristics of each class, depending on the initial composition in the liquid and solid-solid surface energies. Further, it is shown how the surface energy affects the nucleation behavior, in particular of the peritectic phase. Finally we draw conclusions about the range of the possible solid-solid surface energies for which certain morphologies which are experimentally observed can also be simulated and thereafter present an outlook for future work.

## Model Description

A thermodynamically consistent phase-field model is used for the present study of the peritectic reaction in the Fe-C system. The equations are derived from an entropy functional as follows

$$\mathcal{S}(e, c, \phi) = \int_{\Omega} \left( s(e, c, \phi) - \left( \varepsilon a(\phi, \nabla\phi) + \frac{1}{\varepsilon} w(\phi) \right) \right) d\Omega, \quad (3.1)$$

where $e$ is the internal energy of the system, $c = (c_i)_{i=1}^{K}$ is a vector of concentration variables belonging to the $K-1$ dimensional plane, $K$ being the number of components in the system and $\phi = (\phi_\alpha)_{\alpha=1}^{N}$ is a vector of phase-field variables that lies in the $N-1$ dimensional plane, $N$ being the number of phases in the system. $\boldsymbol{\phi}$ and $\boldsymbol{c}$ fulfil the constraints

$$\sum_{i=1}^{K} c_i = 1 \qquad \text{and} \qquad \sum_{\alpha=1}^{N} \phi_\alpha = 1. \tag{3.2}$$

$\varepsilon$ is the small length scale parameter related to the interface width. $s\,(e, \boldsymbol{c}, \boldsymbol{\phi})$ is the bulk entropy density, $a\,(\boldsymbol{\phi}, \nabla\boldsymbol{\phi})$ is the gradient entropy density and $w\,(\boldsymbol{\phi})$ describes the surface entropy potential of the system for pure capillary force driven problems.

We use an obstacle type potential for $w\,(\boldsymbol{\phi})$ of the form,

$$w\,(\boldsymbol{\phi}) = \begin{cases} \dfrac{16}{\pi^2} \sum_{\substack{\alpha,\beta=1 \\ (\alpha<\beta)}}^{N,N} \gamma_{\alpha\beta}\phi_\alpha\phi_\beta + \sum_{\substack{\alpha,\beta,\delta=1 \\ (\alpha<\beta<\delta)}}^{N,N,N} \gamma_{\alpha\beta\delta}\phi_\alpha\phi_\beta\phi_\delta, & \text{if } \boldsymbol{\phi} \in \Sigma \\[2em] \infty, & \text{elsewhere} \end{cases}$$

where $\Sigma = \{\boldsymbol{\phi} \,|\, \sum_{\alpha=1}^{N} \phi_\alpha = 1 \text{ and } \phi_\alpha \geq 0\}$,

$\gamma_{\alpha\beta}$ is the surface entropy density and $\gamma_{\alpha\beta\delta}$ is a term added to maintain the solution at an $\alpha\beta$ interface strictly along the two phase interface.

The gradient entropy density $a\,(\boldsymbol{\phi}, \nabla\boldsymbol{\phi})$ can be written as,

$$a\,(\boldsymbol{\phi}, \nabla\boldsymbol{\phi}) = \sum_{\substack{\alpha,\beta=1 \\ (\alpha<\beta)}}^{N,N} \gamma_{\alpha\beta} \left[ a_c\,(q_{\alpha\beta}) \right]^2 |q_{\alpha\beta}|^2,$$

where $q_{\alpha\beta} = (\phi_\alpha\nabla\phi_\beta - \phi_\beta\nabla\phi_\alpha)$ is a normal vector to the $\alpha\beta$ interface. $a_c\,(q_{\alpha\beta})$ describes the form of the surface energy anisotropy of the evolving phase boundary. For applications to solidification in Fe-C, we use a smooth cubic anisotropy modelled by the expression,

$$a_c\,(q_{\alpha\beta}) = 1 \mp \delta_{\alpha\beta} \left( 3 - 4\frac{|q_{\alpha\beta}|_4^4}{|q_{\alpha\beta}|^4} \right),$$

where $|q_{\alpha\beta}|_4^4 = \sum_i^d (q_{\alpha\beta})_i^4$ and $|q_{\alpha\beta}|^4 = \left[\sum_{i=1}^d (q_{\alpha\beta})_i^2\right]^2$, $d$ being the number of dimensions. $\delta_{\alpha\beta}$ is the strength of the anisotropy. Evolution equations for $c$ and $\phi$ are derived from the entropy functional through conservation laws and phenomenological maximization of entropy, respectively [33, 79]. For an isothermal reaction, the evolution equations for the phase-field variables read:

$$\omega\varepsilon\partial_t\phi_\alpha = \varepsilon\left(\nabla \cdot a_{,\nabla\phi_\alpha}\left(\phi, \nabla\phi\right) - a_{,\phi_\alpha}\left(\phi, \nabla\phi\right)\right) -$$
$$\frac{1}{\varepsilon}w_{,\phi_\alpha}\left(\phi\right) - \frac{f_{,\phi_\alpha}\left(T, c, \phi\right)}{T} + \Xi_\alpha - \Lambda, \tag{3.3}$$

where $\Lambda$ is the Lagrange parameter to maintain the constraint in Eqn. (3.2), $\omega$ is a factor related to the relaxation time constant and $\Xi_\alpha$ is the noise in the evolution equation for the $\alpha$ phase. The formulation contains the notation $\partial_t\phi_\alpha = \partial\phi_\alpha/\partial t$ for the partial derivative of the phase-field variable in time. Further, $a_{,\nabla\phi_\alpha}$, $a_{,\phi_\alpha}$, $w_{,\phi_\alpha}$ and $f_{,\phi_\alpha}$ indicate the derivatives of the respective entropy density with respect to $\nabla\phi_\alpha$ and $\phi_\alpha$. The noise function $\Xi_\alpha$ is such that its amplitude is non-zero only in the liquid and smoothly goes to zero in the bulk solid. The noise amplitude distribution is uniform.

The function $f(T, c, \phi)$ in Eqn.(3.3) describes the Gibbs free energy as a summation of all bulk free energy contributions $f_\alpha(T, c)$ from the phases in the system. For the present investigation, we assume an isothermal condition of the system and use a non-dimensionalized form of the free energies $f_\alpha$ with $RT_p/v_m$ as the energy density scale, where $T_p$ is the peritectic temperature, $R$ is the gas constant and $v_m$ is the molar volume. For simplicity sake, we consider the molar volumes of all the components in all the phases to be equal. We use an ideal solution formulation,

$$f(T, c, \phi) = \sum_{i=1}^K \left(\sum_{\alpha=1}^N c_i L_i^\alpha \frac{(T - T_i^\alpha)}{T_i^\alpha} h_\alpha\left(\phi\right)\right) + T\left(c_i log\left(c_i\right)\right)$$

with,

$$f_\alpha(T, c) = \sum_{i=1}^K c_i L_i^\alpha \frac{(T - T_i^\alpha)}{T_i^\alpha} + T\left(c_i log\left(c_i\right)\right)$$

$$f_l(T, \boldsymbol{c}) = T \sum_{i=1}^{K} \left(c_i \log\left(c_i\right)\right).$$

$f_\alpha(T, \boldsymbol{c})$ is the free energy of the $\alpha$ solid phase. The terms $L_i^\alpha$ and $T_i^\alpha$ are the latent heats and the melting temperatures respectively of the $i^{th}$ component in the $\alpha$ phase. We choose the liquid as the reference state and hence $L_i^\alpha = 0$ where $\alpha$ is the index for the liquid phase in the vector $\boldsymbol{\phi}$. The function $h_\alpha(\boldsymbol{\phi})$ interpolates the free energy of the $\alpha$ phase and we choose it to be of the form $h_\alpha(\boldsymbol{\phi}) = \phi_\alpha^2 (3 - 2\phi_\alpha)$ in the present analysis. In general, other interpolation functions could also be formulated and used which involve other components of the $\boldsymbol{\phi}$ vector. The isothermal non-dimensionalized temperature of the system is denoted by $T$, the scale for the temperature being $T_p$.

The evolution equations for the concentration fields are derived from Eqn. (3.1) and give,

$$\partial_t c_i = \nabla \cdot \left( \sum_{j=1}^{K} M_{ij}\left(\boldsymbol{c}, \boldsymbol{\phi}\right) \nabla \left( \frac{1}{T} \frac{\partial f(T, \boldsymbol{c}, \boldsymbol{\phi})}{\partial c_j} \right) \right)$$

The formulation $L_{ij}(\boldsymbol{c}, \boldsymbol{\phi})$ is capable to describe self and interdiffusion in multicomponent systems. We use the form,

$$M_{ij}(\boldsymbol{c}, \boldsymbol{\phi}) = D_i(\boldsymbol{\phi})c_i \left(\delta_{ij} - c_j\right).$$

The diffusion coefficient is formulated as a linear interpolation across the phases $D_i(\boldsymbol{\phi}) = \sum_{\alpha=1}^{N} D_i^\alpha \phi_\alpha$, where $D_i^\alpha$ is the non-dimensionalized diffusivity of the $i^{th}$ component in the $\alpha$ phase. We use $D^l = D_i^l$ to denote the diffusivity of the components Fe and C in the liquid phase as the reference, where $l$ denotes the liquid phase. The capillary length $d_0 = \sigma / \left(RT/v_m\right)$ with the surface energy $\sigma$, is chosen as the length scale and $d_0^2/D^l$ is the time scale for the simulations.

In the following sections we limit our discussion to $N = 3$ phases and $K = 2$ components; $\boldsymbol{\phi} = (\phi_\delta, \phi_\gamma, \phi_l)$ and $\boldsymbol{c} = (c_{Fe}, c_C)$. The concentration space is one-dimensional and without loss of generality, we define the Fe concentration or $c_{Fe}$ as the independent concentration variable/field.

## Parameters related to the Fe-C phase diagram

The parameters, latent heats and melting temperatures of the free energy $f(T, c, \phi)$ required to fit the phase diagram of the Fe-C system in the vicinity of the peritectic temperature are listed in the following two ($2 \times 2$) matrices.

$$
L_i^\alpha := \left( \begin{array}{ccc}
 & Fe & C \\
\gamma & 0.76382 & 4.80548 \\
\delta & 0.84156 & 4.80548
\end{array} \right)
$$

$$
T_i^\alpha := \left( \begin{array}{ccc}
 & Fe & C \\
\gamma & 1.02193 & 0.81242 \\
\delta & 1.0244 & 0.73635
\end{array} \right)
$$

The parameters are determined such that the compositions of the solid and liquid phases at the peritectic temperature as well as the slope of the $\gamma$-liquidus close to the peritectic temperature are accurately reproduced. The phase diagram containing the solidus and liquidus lines, their stable and metastable extensions are displayed in Fig. 3.1.

## Surface energy calculations

The surface energies $\tilde{\sigma}_{\delta l}$ and $\tilde{\sigma}_{\gamma l}$ of the $\delta$-ferrite/liquid and of the $\gamma$-austenite/liquid phase boundaries are known to be equal. We use the surface energies in the length scale of the non-dimensionalization. However, there is an excess contribution from the chemical free energies which needs to be included, if the interface between the two phases does not strictly evolve along the co-existence line. We calibrate the surface entropy densities $\gamma_{\delta l}$ and $\gamma_{\gamma l}$ such that the final non-dimensional values of the surface energies are $\tilde{\sigma}_{\delta l} = \tilde{\sigma}_{\gamma l} = 1.0$. The excess chemical free energy is calculated using the following procedure. Each solid phase is allowed to equilibrate with the liquid forming a planar interface, by setting the temperature of the system at the peritectic point and the compositions of the solid and the liquid phases as the equilibrium compositions. The surface energy excess is calculated including the excess of the grand chemical potential $f(T, c, \phi) - \mu_{Fe} c_{Fe}$ of the final equilibrated structure with respect to that of any of the bulk phases. Since we set the bulk compositions at the values from the liquidus and solidus lines of the phase diagram, the grand

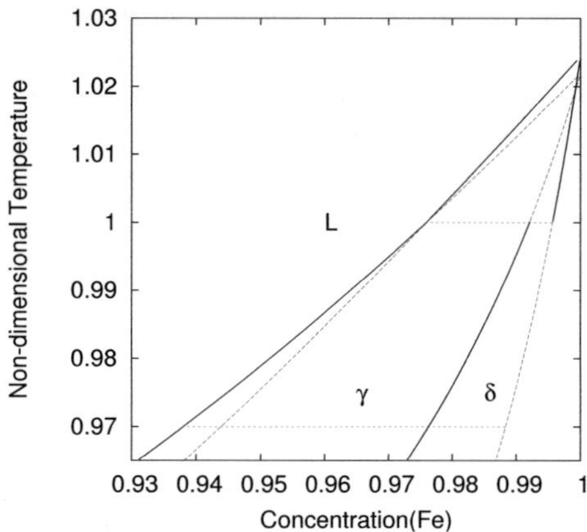


**Figure 3.1.:** Phase diagram of the Fe-C system close to the peritectic reaction showing stable (solid lines) and metastable solidus and liquidus lines (dashed-lines). The tie-lines at the temperature $T = 1.0$ and $0.97$ denote the temperature $T_p$ of the peritectic reaction and the temperature at which the system is undercooled and at which the simulations were performed respectively.

chemical potential of the phases at these compositions are equal. The surface energy excess is given by the relation,

$$\tilde{\sigma}_{\alpha l} \;\; = \;\; \int_X \left( T_p \varepsilon a \left( \phi, \nabla \phi \right) + \frac{T_p}{\varepsilon} w \left( \phi \right) + \Delta \Psi(T, \boldsymbol{c}, \phi) \right) dX$$

with,

$$\Delta \Psi(T, \boldsymbol{c}, \phi) \;\; = \;\; f \left( T_p, \boldsymbol{c}, \phi \right) - f_l - \mu_{Fe}(T_p) \left( c_{Fe} - c^l_{Fe} \right),$$

where $\mu_{Fe}\left( T_p \right) = \dfrac{\partial f \left( T, \boldsymbol{c}, \phi \right)}{\partial c_{Fe}}$ is the equilibrium chemical potential that is established during the equilibration of the planar interface between the phases. The partial derivative are taken respecting the constraint of the concentration fields Eqn. (3.2). $c_{Fe}$ is the concentration profile that stabilizes between the phases, where stabilization implies that, both the concentration and phase-field profiles are stationary. $c^l_{Fe}$ is the concentration of Fe in the bulk liquid. The surface energies of the solid-solid interfaces are also calculated in a similar manner, the equilibration being done between the two solid phases. Since the grand chemical potential of all the three phases $\gamma$, $\delta$, $l$ is the same at the peritectic temperature, the liquid can be used as the reference to calculate the surface energy excess.

## 3.2. Results and Discussion

The evolution equations for the phase-field and concentration variables are numerically discretized using an explicit forward in time, finite difference scheme. The resulting discrete set of equations is solved applying a parallel solver depending on the MPI (message parsing interface) standard. For this, the domain is decomposed in one dimension in equal parts for each node used for the simulation. Due to the fact that the phase-field equation is only solved in the diffuse interface between two phase fields, the calculation time of each node can be different. To optimize time and resources, a dynamics domain decomposition algorithm is used for the redistribution of layers during runtime depending on the calculation time of each node. In the present work we consider three types of growth morphologies:

- engulfing of the properitectic $\delta$-phase by the peritectic $\gamma$-phase

- primary dendritic growth with nucleation events and

- growth of the $\gamma$-phase along a plate substrate.

Table 3.1 summarizes the parameters used for simulating the described morphologies.

**Table 3.1.:** Numerical parameters used for the phase-field simulations. $Nx, Ny, Nz$ denote the domain dimensions, $\Delta X$ denotes the grid spacing, $\Delta t$, the time step width, and $\varepsilon$ the interface width

| Parameters | Dendrite-2D Fig.3.6 | Dendrite-3D Fig. 3.7 | Engulf-2D Fig. 3.2 | Engulf-3D Fig. 3.5 | Plate-2D Fig. 3.9 |
|---|---|---|---|---|---|
| $Nx \times Ny \times Nz$ | $1000 \times 1000$ | $500 \times 500 \times 500$ | $500 \times 500$ | $500 \times 500 \times 500$ | $200 \times 500$ |
| $\Delta X$ | 10 | 20 | 10 | 10 | 1 |
| $\Delta t$ | 7.5 | 2.5 | 5 | 5 | 0.08 |
| $\varepsilon$ | 80 | 80 | 80 | 80 | 8 |

## 3.2.1. Engulfing of the pro-peritectic phase by the peritectic phase

A growth morphology observed in the Fe-C system is the engulfing microstructure. Controlled by diffusion in the liquid, the peritectic $\gamma$-phase grows over the properitectic $\delta$-phase until complete engulfment. The subsequent process is a phase transformation ($\delta \to \gamma$) taking place on a longer time scale, as it is solely driven by diffusion in the solid. The phase transition kinetics is dependent on the value of the solid-solid surface energies and the concentration of the liquid from which the two solids, the $\gamma$- and $\delta$-phase, are evolving. While the solid-liquid surface energies are known fairly accurately for the $\gamma$- and the $\delta$-phases, the solid-solid surface energy $\tilde{\sigma}_{\gamma\delta}$ is unknown. The phase-field simulations are employed to locate the range of $\tilde{\sigma}_{\gamma\delta}$ and to investigate the effect of the concentration in the liquid on the engulfing behavior of the $\gamma$-phase. The parametric study is conducted at a non-dimensionalized temperature T $= 0.97$ with a non-dimensional diffusivity $D^l_{Fe} = 1.0$ in the liquid and $D^s_{Fe} = 0.01$ in the solid. Fig. 3.2 shows the volume change (in 2D: height in non-growth direction is 1) of the peritectic $\gamma$-phase in time for three different initial Fe concentrations of the liquid, $c^l_{Fe} = 0.95$, $0.96$ and $0.97$. The points on the graph are plotted until the properitectic $\delta$-phase is totally engulfed by

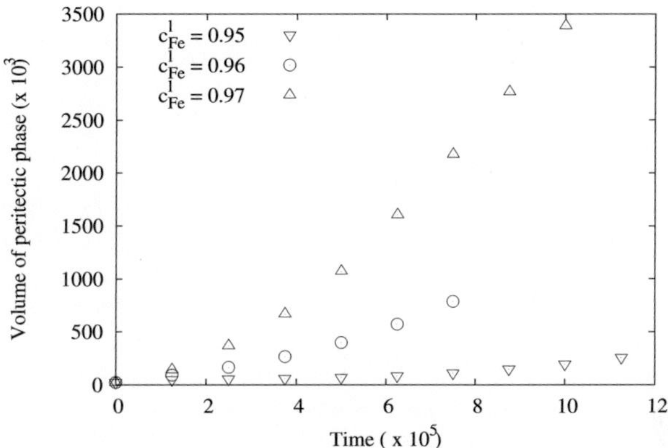


**Figure 3.2.:** Growth of the peritectic $\gamma$-phase during engulfment, for different concentrations $c^l_{Fe}$ of the liquid phase and for a constant surface energy $\tilde{\sigma}_{\gamma\delta} = 0.33$

the peritectic $\gamma$-phase. The volume of the $\gamma$-phase at a particular time, increases for higher values of $c^l_{Fe}$. Investigations for different values of the solid-solid surface energy $\tilde{\sigma}_{\gamma\delta}$, revealed reduction in the rate of engulfment of the $\delta$-phase with increase in the solid-solid surface energy, for a particular value of $c^l_{Fe}$. Fig. 3.3 displays the comparison between the rate of volume change $\partial V/\partial t$ of the peritectic and properitectic phase, as a function of time, for the liquid concentrations $c^l_{Fe} = 0.95$ and $c^l_{Fe} = 0.97$ and for the surface energy $\tilde{\sigma}_{\gamma\delta} = 0.33$. We see two different types of behaviour at the two levels of super-saturation. At the lower supersaturation Fig. 3.3a, there is an initial decrease in the shrinking rate of the $\delta$-phase along with a corresponding decrease in the rate of growth of the $\gamma$-phase. This behaviour ends with the attainment of the minimum in the reduction rate of the $\delta$-phase beyond which, there is increase in the solidification rate of the $\gamma$-phase. We note that the behaviour of the rates of evolution of the solid phases are always of opposite nature, which implies, that the solute transfer at the small-scale, from one solid to the other through the liquid as the transport medium, is the principal growth limiting process, rather than the long range diffusion in the liquid. This is because, individually,

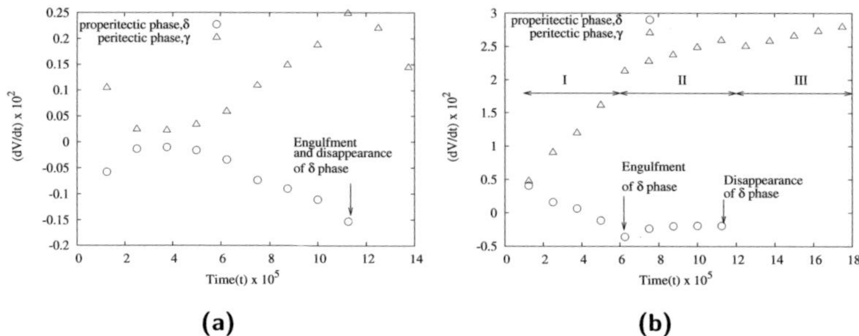


**Figure 3.3.:** Rate of volume change of the peritectic and properitectic phases with time for a constant surface energy $\tilde{\sigma}_{\gamma\delta} = 0.33$ and for two different liquid concentrations (a) $c^l_{Fe} = 0.95$ and (b) $c^l_{Fe} = 0.97$.

each phase is set above the critical size, beyond which growth occurs in the liquid. When set in a coupled configuration, the growth rate of the $\gamma$-phase is higher, resulting from the larger super-saturation of this phase, as is evident from the phase diagram. This causes the region local to the $\delta$-phase to become enriched in C, which reduces the effective super-saturation for the $\delta$-phase and causes it to shrink. Thus, the $\delta$- phase acts as a sink of C atoms by melting. To understand the growth process in full detail, one would need to construct the free boundary problem for the present configuration and study the analytical solutions, which is not in the scope of the present chapter. Simulations, however give a reasonable insight into the growth dynamics. At this concentration, the point of engulfment of the $\delta$-phase coincides with the disappearance of the phase. This also shows that, at this value of surface tension, the concentration of the liquid $c^l_{Fe} = 0.95$ is the minimum requirement to observe this microstructure. Also, we notice the rate of change of the $\gamma$-phase to reduce, after the point of engulfment, from which we can again infer, that the small scale diffusion near the triple point, is the prominent growth mode for the $\gamma$-phase, during the process of engulfment. At the higher super-saturation corresponding to the concentration, $c^l_{Fe} = 0.97$, Fig. 3.3b, we have three growth regimes possible. In the region (I), before the engulfment of the $\delta$-phase we again notice opposite nature of change of the growth rates for the solid phases, as we did for the case of the lower super-saturation. After engulfment (region II), the reduction rate of the $\delta$-phase approximately becomes a constant, while the evolution rate of the $\gamma$-phase shifts to a linear curve of reduced slope. Comparing the growth dynamics in (region II) with the evolution characteristics after the disappearance of the $\delta$-phase (region III), shows that the long range diffusion in the liquid is the principal growth mechanism after the engulfment, as the there is little change in the slope of the variation in the evolution rate. Based on the parametric studies evaluating the influence of $c_{Fe}^l$ on the engulfing morphology, we analysed the effect of the surface energy $\tilde{\sigma}_{\gamma\delta}$. Fig. 3.4 compares the dynamics of the phase transformations for the different surface energies $\tilde{\sigma}_{\gamma\delta} = 0.33$, 0.36, 0.39, 0.42 and 0.45. The points are plotted until one of the $\gamma$- or $\delta$-phases

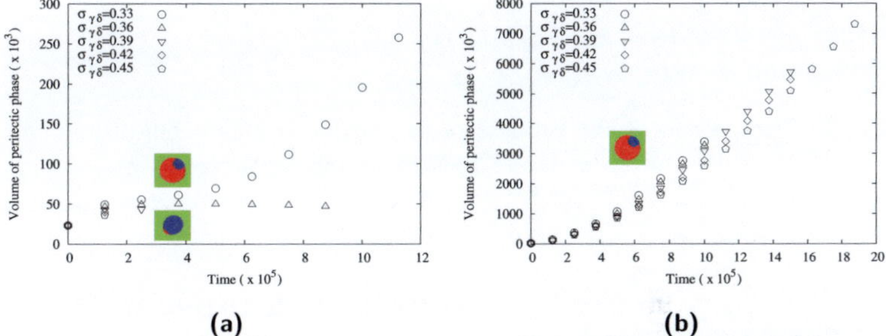

(a)                                                (b)

**Figure 3.4.:** Volume change of the peritectic $\gamma$-phase in time, illustrating the influence of the surface energy $\tilde{\sigma}_{\gamma\delta}$ on the type of morphology and on the growth rate for two different concentrations (a) $c_{Fe}^l = 0.95$ and (b) $c_{Fe}^l = 0.97$
(Please read the legends in the graphs $\sigma$ as $\tilde{\sigma}$)

disappears. For the liquid concentration at $c_{Fe}^l = 0.95$ (Fig. 3.4a), the peritectic $\gamma$-phase engulfs the properitectic phase for $\tilde{\sigma}_{\gamma\delta} = 0.33$, while for other values of surface energies, the contrary happens. Setting $c_{Fe}^l = 0.97$ (Fig. 3.4b), the $\delta$-phase becomes completely embedded for all values of $\tilde{\sigma}_{\gamma\delta}$. Moving towards higher values of $\tilde{\sigma}_{\gamma\delta}$ leads to lesser volume of the engulfed $\delta$-phase. To conclude: The observation of engulfing morphologies at this temperature and concentration requires surface energies $\tilde{\sigma}_{\gamma\delta}$ in the considered range. The volume of the $\gamma$-phase increases to much larger values if the liquid concentration changes from $c_{Fe}^l = 0.95$ to higher values, as in Fig. 3.4b. Engulfing microstructures were also obtained in 3D, as exemplarily shown in Fig. 3.5, for simulation data set at a non-dimensional temperature $T = 0.97$, $c^l_{Fe} = 0.96$ and $\tilde{\sigma}_{\gamma\delta} = 0.39$.

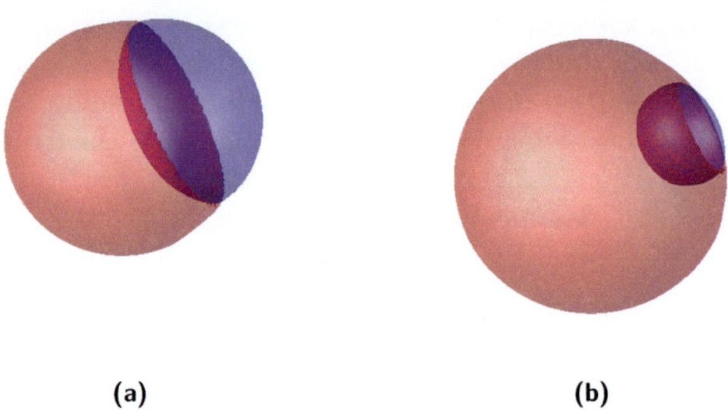

(a)              (b)

**Figure 3.5.:** Simulation of the process of properitectic phase engulfment in 3D at (a) an early stage and (b) a late stage of the peritectic reaction.

## 3.2.2. Primary dendritic growth with nucleation events

The peritectic reaction and growth of $\gamma$-phase on the periphery of a $\delta$-phase dendrite was investigated by simulations previously in [113]. However, the location of first nucleation and subsequent growth of $\gamma$-phase has not been the object of consideration so far. We address this issue and perform 2D and 3D phase-field simulations, to determine heterogeneous nucleation sites and solidifying microstructures, depending on the surface energies $\tilde{\sigma}_{\gamma\delta}$. As illustrated in Fig. 3.6a, heterogeneous nucleation events of the $\gamma$-phase on top of the dendritic tips occurs for $\tilde{\sigma}_{\gamma\delta} = 0.39$, along with homogeneous nucleation in the undercooled liquid for the chosen super-saturation. For higher surface energy $\tilde{\sigma}_{\gamma\delta} = 0.778$, no heterogeneous nucleation on the dendritic surface can be observed, Fig. 3.6b. Under these conditions, the two solid phases only nucleate homogeneously in the liquid. According to Eqn. 3.3, nucleation is effected through uniform noise in the liquid, similar to the formulation of Granazy et al. [95, 131]. The first nucleation event

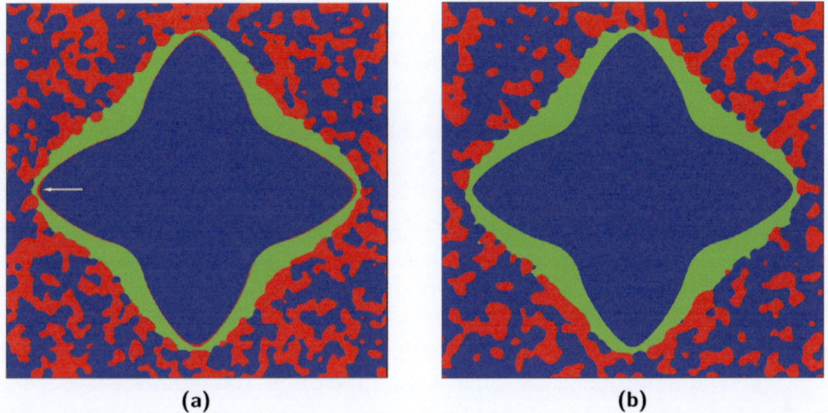

**Figure 3.6.:** Phase-field simulations of nucleation in the peritectic Fe-C system for two different surface energies $\tilde{\sigma}_{\gamma\delta} = 0.39$ and $\tilde{\sigma}_{\gamma\delta} = 0.778$ in comparison. (a) For the lower value of the solid-solid surface energy, heterogeneous nucleation can be seen at the dendrite tips, (b) For the higher value, the nuclei are purely formed homogeneously in the liquid. Color coding; Red: $\gamma$- phase, Blue: $\delta$-phase, Green: Liquid

occurs at the place in the domain, where the required critical fluctuation for nucleation is the smallest. The closer $c^l_{Fe}$ approaches the value of the solidus, the smaller is the magnitude of the critical fluctuation, because the driving force for solidification increases. It follows from classical nucleation theory that the free energy size of the critical fluctuation varies inversely to the supersaturation: which increases as we move towards the solidus. It is important however, to note that the driving force is the difference of the bulk free energies of the transforming phases. In the case of phase-field simulations however, we have the possibility of critical nuclei being sub-solid i.e. consisting of only the interface. At concentrations near the solidus line the critical nuclei become smaller following the increase in supersaturation. They loose their bulk properties and importantly because of lesser volume have a small barrier to nucleation. Close to the properitectic dendrite, the concentration is enriched in C (deficient in Fe), compared to the concentration in the liquid far away from the interface. The concentration around the dendrite is further away from the solidus of the $\gamma$-phase and hence, the critical fluctuation to nucleate the $\gamma$-phase is larger than away from the dendrite. As a result the nucleation events occur in the liquid phase at a distance from the dendrite. The simulation results for the higher value of the surface energy $\tilde{\sigma}_{\gamma\delta}$ agrees with the drawn explanation. Heterogeneous nucleation becomes a possibility as $\tilde{\sigma}_{\gamma\delta}$ is reduced. The heterogeneous nucleation appears at the dendrite tip, where the concentration of the liquid is poorer in C compared to the concave part of the dendrite. A similar nucleation behaviour can be observed in a 3D simulation of an evolving $\delta$-dendrite with a cubic crystal symmetry followed by nucleation of the $\gamma$-phase, (Fig. 3.7a and 3.7b). The $\gamma$-phase nucleates on the edges and tips of the dendrite, whereas no nucleation is observed in the concave regions of higher C concentration, compared to the tips and edges. The reasoning is the same as in the 2D simulations discussed above. Fig. 3.7a shows an evolving dendrite with four fold weak cubic anisotropy. In addition to the images in Fig. 3.7a and Fig. 3.7b highlighting the regions occupied by the $\delta$- and $\gamma$-phases, Fig. 3.7c is the Fe concentration, mapped on the iso-surface at $\phi_\delta = 0.5$ of the properitectic dendrite. Comparing Fig. 3.7b and Fig. 3.7c, accentuates the point, that the nucleation occurs at places where the barrier to nucleation is the lowest. For the chosen parameter set and noise amplitude, no homogeneous nucleation occurs.

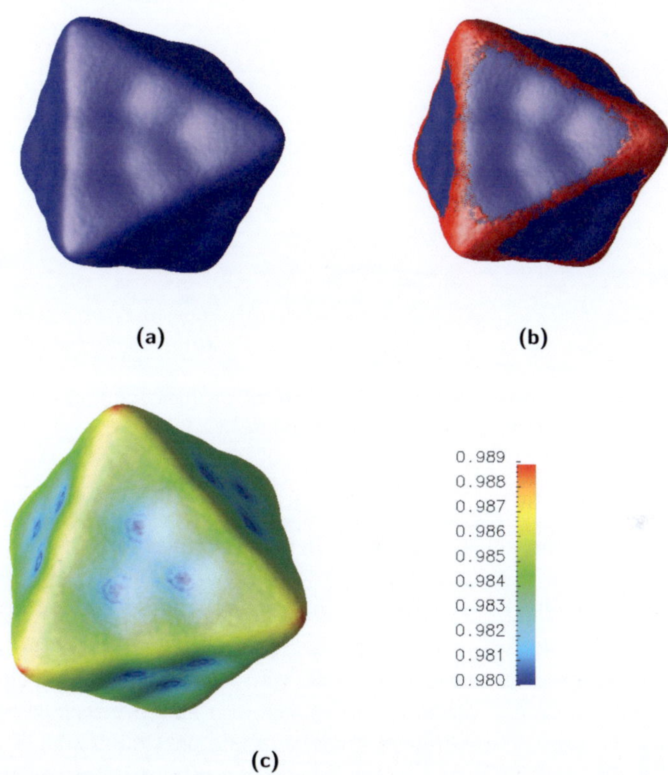

**Figure 3.7.:** Phase-field simulation of a dendritic microstructure with four-fold surface energy anisotropy: (a) final shape of the properitectic phase, (b) heterogeneous nucleation at the edges and tips of the dendrite and (c) concentration of Fe at the dendritic surface for the same time step as in (b).

We would like to point out that, we do not expect spurious effects like the stabilization of meta-stable phases due to the reduction in the height of the surface potential for the interface thickness that we choose. This is because our driving forces are of very low magnitude in the simulations. Fig. 3.8a and Fig. 3.8b make this point clear. We calculated the critical

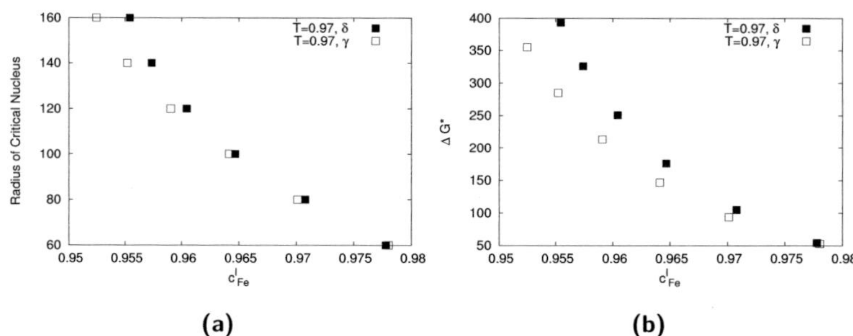

(a)        (b)

**Figure 3.8.:** Calculation of the radius of the critical profiles for homogeneous nucleation at T=0.97 in a) and barrier to nucleation in b).

phase field profiles of the solid phases in equilibrium with the liquid at various super-saturations, by solving the Euler-Lagrange equations $\dfrac{\delta S}{\delta \phi} = 0$ and $\dfrac{\delta S}{\delta c} = constant$ as in [95, 131]. From this, we calculated the barrier to homogeneous nucleation $\Delta G^*$ as the excess grand chemical potential and inclusive of all the surface excesses, Fig. 3.8b, and the radius of the critical nucleus Fig. 3.8a, which corresponds to the size of a sharp interface solid, which equals the volume of the critical nucleus of the particular phase. We see that, at the concentration of $c^l_{Fe} = 0.97$ we have the critical radius to be 80 times the capillary length, which shows that the effect of the driving forces becomes comparable to the capillary forces due to the surface energy at these sizes. Although we have considered solid-liquid equilibrium, we expect, the solid-solid transformation energies to be of the same order or smaller, because of the similarity of the solid-liquid equilibrium among the $\delta$- and $\gamma$-phases. To be able to simulate large microstructures then, within manageable number of grid points, it is necessary to choose the interface widths of the order we have chosen and as is also customary in the phase field method, to use interface widths of the order of the smallest principal length scale that one is trying to resolve. However, one must take care, that the model used for this study, does not contain the thin interface corrections, [46], requisite in large interface simulations for quantitative results, and hence our kinetics will be off by some order. In light of this, the results, only qualitatively show the influence of the surface energy and super-saturation on the kinetics of transformation. However, since we were able to treat our surface excesses correctly, our thermodynamic predictions regarding nucleation sites and range of surface energies are reliable.

### 3.2.3. Growth of the peritectic phase along a plate substrate

During the peritectic reaction $L + \delta \to \gamma$, the triple junction formed by the three phases drives the phase transitions. A well-known morphology is the propagation of the peritectic $\gamma$-phase on top of the substrate, consisting of properitectic $\delta$-phase, Fig. 3.9. The process is liquid-diffusion controlled.

**Figure 3.9.:** The peritectic $\gamma$-solid growth on top of a properitectic $\delta$-substrate. The dynamics of dissolution is driven by diffusion of solute in the liquid phase.

The concentration of the liquid is chosen such that the $\delta$-phase is in equilibrium at a non-dimensionalized temperature of 0.97. The large substrate of the properitectic phase ensures that curvature undercooling of the solid-liquid interface is negligible.

## 3.3. Conclusion

We use a phase-field model to investigate the evolution of different growth morphologies during the peritectic reaction in the Fe-C system through 2D and 3D simulations. Predictions of the solid-solid surface energies $\tilde{\sigma}_{\gamma\delta}$ are made based on comparisons between experimental observations, previous simulations and the current work. We discuss the effect of $\tilde{\sigma}_{\gamma\delta}$ on nucleation behavior. We find heterogeneous nucleation on a primary dendrite, strongly depending on the solid-solid surface energy. Based on our study, we propose that the value for $\tilde{\sigma}_{\gamma\delta}$ should lie in a range causing the two solids, the $\gamma$- and $\delta$-phase, to form low contact angles $< 26°$ (illustration in Fig. 3.10). Furthermore, we observe heterogeneous nucleation becoming a possibility

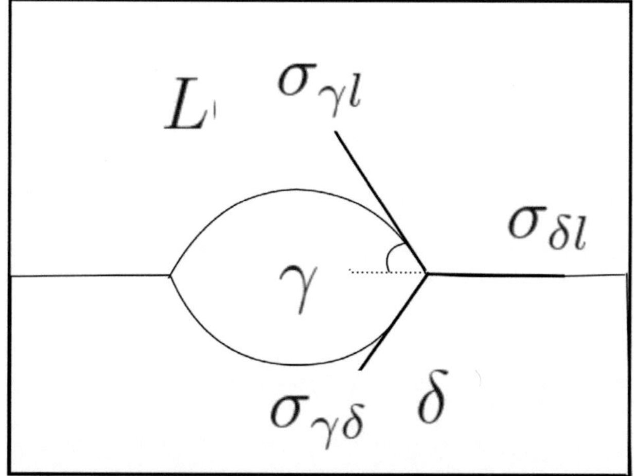


**Figure 3.10.:** Illustration of the angle that the surface tangent at the $l-\delta$ interface close to the triple junction, makes with the horizontal (original substrate).

for such values of $\tilde{\sigma}_{\gamma\delta}$. The main emphasis of our present work is to understand the role of $\tilde{\sigma}_{\gamma\delta}$ on the growth morphologies and on nucleation sites. Hence other forms of free energy such as from thermodynamic databases, e.g. CALPHAD, were not used in order to simplify the analysis. Future work would involve a more rigorous treatment of nucleation behavior to predict nucleation sites on evolving substrates and composition domains based on a careful study of critical nuclei.

# Chapter 4

# Study of three phase growth in ternary alloys

## 4.1. Introduction

Eutectic alloys are of major industrial importance because of their low melting points and their interesting mechanical properties. They are also interesting for physicists because of their ability to form a large variety of complex patterns, which makes eutectic solidification an excellent model system for the study of numerous nonlinear phenomena.

In a binary eutectic alloy, two distinct solid phases co-exist with the liquid at the eutectic point characterized by the eutectic temperature $T_{\mathrm{E}}$ and the eutectic concentration $C_{\mathrm{E}}$. If the global sample concentration is close to the eutectic concentration, solidification generally results in composite patterns: alternating lamellae of the two solids, or rods of one solid immersed in a matrix of the other, grow simultaneously from the liquid. The fundamental understanding of this pattern-formation process was established by Jackson and Hunt (JH) [44]. They calculated approximate solutions for spatially periodic lamellae and rods that grow at constant velocity $v$, and established that the average front undercooling, that is, the difference between the average front temperature and the eutectic temperature, follows the relation

$$\Delta T = K_1 v \lambda + \frac{K_2}{\lambda}, \tag{4.1}$$

where $\lambda$ is the width of one lamella pair (or the distance between two rod centers), $v$ is the velocity of the solidification front, and $K_1$ and $K_2$ are constants whose value depends on the volume fractions of the two solid phases and various materials parameters [44]. The two contributions in Eq. (4.1) arise from the redistribution of solute by diffusion through the liquid and the curvature of the solid-liquid interfaces, respectively.

The front undercooling is minimal for a characteristic spacing

$$\lambda_{JH} = \sqrt{\frac{K_2}{K_1 v}}. \tag{4.2}$$

The spacings found in experiments in massive samples are usually distributed in a narrow range around $\lambda_{JH}$ [116]. However, other spacings can be reached in directional solidification experiments by imposing a solidification velocity that varies with time. In this way, the stability of steady-state growth can be probed [34]. In agreement with theoretical expections [73], steady-state growth is stable over a range of spacings that is limited by the occurrence of dynamic instabilities. For low spacings, a large-scale lamella (or rod) elimination instability is observed [5]. For high spacings, the type of instability that can be observed depends on the sample geometry. For thin samples, various oscillatory instabilities and a tilt instability can occur, depending on the alloy phase diagram and the sample concentration. Beyond the onset of these instabilities, stable tilted patterns as well as oscillatory limit cycles can be observed in both experiments and simulations [34, 51]. For massive samples, a zig-zag instability occurs for lamellar eutectics [2, 85], whereas rods exhibit a shape instability [86].

In summary, pattern formation in binary eutectics is fairly well understood. However, most materials of practical importance have more than two components. Therefore, eutectic solidification in multicomponent alloys has received increasing attention in recent years. A particularly interesting situation arises in alloy systems that exhibit a ternary eutectic point, at which four phases (three solids and the liquid) coexist. At such a quadruple point, three binary "eutectic valleys", that is, monovariant lines of three-phase coexistence, meet. The existence of three solid phases implies that there is a far greater variety of possible structures, even in thin samples. Indeed, for two solids $\alpha$ and $\beta$, an array $\alpha\beta\alpha\beta\ldots$ is the only possibility for a composite pattern in a thin sample; the only remaining degree of freedom is the spacing. With an additional $\gamma$ solid, an infinite number of distinct periodic cycles with different sequences of phases are possible. The simplest cycles are $\alpha\beta\gamma\alpha\beta\gamma\ldots$ and $\alpha\beta\alpha\gamma\alpha\beta\alpha\gamma\ldots$ and permutations. Clearly, cycles of arbitrary length, and even non-periodic configurations are possible. An interesting question is then which configurations, if any, will be favored.

In preliminary works, the occurrence of lamellar structures has been reported in experiments in massive samples [9, 23, 42, 52, 74, 102, 104]. The spatio-temporal evolution in ternary eutectic systems was observed in thin samples (quasi-2D experiments) in both metallic [101] and organic systems [139]. In both cases, the simultaneous growth of three distinct solid phases from the liquid with a $(\alpha\beta\alpha\gamma)$, (named ABAC in Ref. [139])

stacking was observed. Measurements in both cases revealed that $\lambda^2 v$ was approximately constant, in agreement with the JH scaling of Eq. (4.2).

On the theoretical side, models that extend the JH analysis from binary to ternary eutectics for three different growth morphologies (rods and hexagon, lamellar, and semi-regular brick structures) were proposed by Himemiya *et al.* [41]. The relation between front undercooling and spacing is still of the form given by Eq. (4.1), with constants $K_1$ and $K_2$ that depend on the morphology. The differences between the minimal undercoolings for different morphologies were found to be small. No direct comparison to experiments was given.

Finally, ternary eutectic growth has also been investigated by phase-field methods in Refs. [7, 40], who have studied different stacking sequences formed by $\alpha = Ag_2Al$, $\beta = (\alpha\ Al)$ and $\gamma = Al_2Cu$ in the ternary system Al-Cu-Ag, while transients in the ternary eutectic solidification of a transparent In-Bi-Sn alloy were studied both by phase field modeling and experiments [101].

The purpose of the present chapter is to carry out a more systematic investigation of lamellar ternary eutectic growth. The main questions we wish to address are (i) can an extension of the JH theory adequately describe the properties of ternary lamellar arrays and reveal the differences between cycles of different stacking sequences, and (ii) what are the instabilities that can occur in such patterns. To answer these questions, we develop a generalization of the JH theory to ternary eutectics which is capable of describing the front undercoolings of periodic lamellar arrays with arbitrary stacking sequence. Its predictions are systematically compared to phase-field simulations. We use a generic thermodynamically consistent phase-field model [33, 79]. While this model is known to exhibit several thin-interface effects which limit its accuracy [6, 30, 46, 48, 55], we show here that we can obtain a very satisfying agreement between theory and simulations if the solid-liquid interfacial free energy is evaluated numerically. In particular, the minimum-undercooling spacings are accurately reproduced for all stacking sequences that we have simulated.

The model is then used to systematically investigate the instabilities of lamellar arrays, in particular for large spacings. We find that, as for binary eutectics, the symmetry elements of the steady-state array determine the possible instability modes. Whereas the calculation of a complete stability diagram is not feasible due to the large number of independent parameters, we find and characterize several new instability modes. Besides these oscillatory modes that are direct analogs of the ones observed in binary eutectics, we also find a new type of instability which occurs at small spacings: cycles in which the same phase appears more than once can undergo an instability during which one of these lamellae is eliminated; the system therefore transits to a different (simpler) cycle. Furthermore, we also find that the occurrence of this type of instability can be well predicted by our generalized JH theory.

The remainder of the chapter is organized as follows. In Sec. 4.2, we develop the generalized JH theory for ternary eutectics and calculate the undercooling-spacing relationships for several simple cycles. In Sec. 4.3, the phase-field model is outlined and its parameters are related to the ones of the theory. Sec. 4.4 presents the simulation results concerning both steady-state growth and its instabilities. In Sec. 4.5, we briefly discuss questions related to pattern selection and present some preliminary simulations in three dimensions. Sec. 4.6 concludes the chapter.

## 4.2. Theory

We consider a ternary alloy system consisting of components $A, B$ and $C$, which can form three solid phases $\alpha, \beta$, and $\gamma$ upon solidification from the liquid $l$. The concentrations of the components (in molar fractions) are denoted by $c_A, c_B$ and $c_C$ and fulfill the constraint

$$c_A + c_B + c_C = 1. \tag{4.3}$$

This obviously implies that there are only two independent concentration fields.

As is customary, isothermal sections of the ternary phase diagram can be conveniently displayed in the Gibbs simplex. We are interested in alloy systems that exhibit a ternary eutectic point: four-phase coexistence between three solids and the liquid. The isothermal cross-section at the ternary eutectic temperature is displayed in Figure 4.1, here for the particular example of a completely symmetric phase diagram.

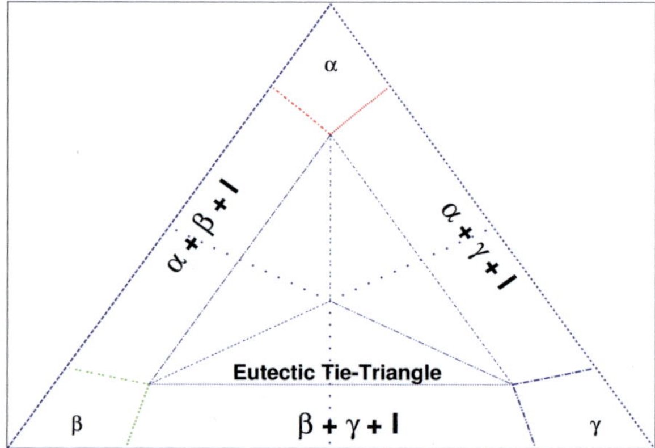


**Figure 4.1.:** Projection of the ternary phase diagram for a model symmetric ternary eutectic system on the Gibbs simplex. The triangle at the center is the tie-triangle at the eutectic temperature where four phases $\alpha, \beta, \gamma$, and $l$ are in equilibrium. The diagram also contains the information on three-phase equilibria. The liquidus lines corresponding to each of these equilibria ("eutectic valleys") are shown by dotted lines which meet at the center of the simplex, which is also the concentration of the liquid at which all the three solid phases and the liquid are at equilibrium.

The concentration of the liquid is located in the center of the simplex ($c_A = c_B = c_C = 1/3$), and the three solid phases are located at the corners of the eutectic tie triangle. For higher temperatures, no four-phase coexistence is possible, but each pair of solid phases can coexist with the liquid (three-phase coexistence). Each of these three-phase equilibria is a eutectic, and the loci of the liquid concentrations at three-phase coexistence as a function of temperature form three "eutectic valleys" that meet at the ternary eutectic point. On each of the sides of the simplex (with the temperature as additional axis), a binary eutectic phase diagram is found.

The key point for the following analysis is the temperature of solid-liquid interfaces, which depends on the liquid concentration, the interface curvature, and the interface velocity. The dependence on the concentration is described by the liquidus surface, which is a two-dimensional surface over the Gibbs simplex. This surface can hence be characterized by two independent liquidus slopes at each point. For each phase $\nu$ ($\nu = \alpha, \beta, \gamma$), we choose the two liquidus slopes with respect to the minority components. Thus, for the $\alpha$ phase, the interface temperature is given by the generalized Gibbs-Thomson relation,

$$T^\alpha_{\text{int}} - T_{\text{E}} = m^\alpha_B(c_B - c^E_B) + m^\alpha_C(c_C - c^E_C) - \Gamma_\alpha \kappa - \frac{v_n}{\mu^\alpha_{\text{int}}}, \qquad (4.4)$$

where $c_B$ and $c_C$ are the concentrations in the liquid adjacent to the interface, $c^E_B$ and $c^E_C$ their values at the ternary eutectic point, and $m^\alpha_B = \left.\frac{dT_\alpha}{dc_B}\right|_{c_C=\text{const}}$ and $m^\alpha_C = \left.\frac{dT_\alpha}{dc_C}\right|_{c_B=\text{const}}$ the liquidus slopes taken at the ternary eutectic point. Furthermore, $\Gamma_\alpha = \tilde{\sigma}_{\alpha l} T_{\text{E}}/L_\alpha$ is the Gibbs-Thomson coefficient, with $\tilde{\sigma}_{\alpha l}$ the solid-liquid surface tension and $L_\alpha$ the latent heat of fusion per unit volume, and $\mu^\alpha_{\text{int}}$ is the mobility of the $\alpha$-liquid interface. For the typical (slow) growth velocities that can be attained in directional solidification experiments, the last term, which represents the kinetic undercooling of the interface, is very small. It will therefore be neglected in the following. The expression for the other solid phases are obtained by cyclic permutation of the indices.

In the spirit of the original Jackson-Hunt analysis, for the calculation of the diffusion field in the liquid, the concentration differences between solid and liquid phases are assumed to be constant and equal to their values at the ternary eutectic point. Since we are interested in ternary coupled growth, which will take place at temperatures close to $T_E$, this should be a good approximation. Thus, we define

$$\Delta c_j^\nu = c_j^l - c_j^\nu \qquad \text{with} \qquad j = A, B, C \qquad \text{and} \qquad \nu = \alpha, \beta, \gamma.$$

In this approximation, the Stefan condition at a $\nu$-$l$ interface, which expresses mass conservation upon solidification, reads

$$\partial_n c_j = -\frac{v_n}{D} \Delta c_j^\nu, \tag{4.5}$$

where $\partial_n c_j$ denotes the partial derivative of $c_j$ in the direction normal to the interface, $v_n$ is the normal velocity of the interface (positive for a growing solid), and $D$ is the chemical diffusion coefficient, for simplicity assumed to be equal for all the components.

We consider a general periodic lamellar array with $M$ repeating units consisting of phases $(\nu_0, \nu_1, \nu_2, \ldots, \nu_{M-1})$ where each $\nu_i$ represents the name of one solid phase $(\alpha, \beta, \gamma)$ in the sequence, with a repeat distance (lamellar spacing) $\lambda$. The width of the $j$-th single solid phase region is $(x_j - x_{j-1})\lambda$, with $x_0 = 0$ and $x_M = 1$, and the sum of all the widths corresponding to any given phase is its volume fraction $\eta_\nu$. The eutectic front is assumed to grow in the $z$ direction with a constant velocity $v$.

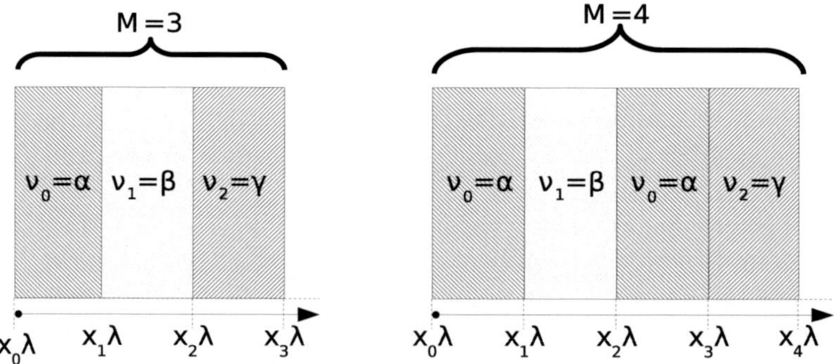


**Figure 4.2.:** Two examples for periodic lamellar arrays with $M = 3$ and $M = 4$ units.

## 4.2.1. Concentration fields

First, we consider the diffusion fields of the components $A, B, C$ ahead of a growing eutectic front. For the calculation of the concentration fields, the front is supposed to be planar, as in the sketches of Figure 4.2. We make the following Fourier series expansion for $c_A$ and $c_B$

$$c_X = \sum_{n=-\infty}^{\infty} X_n e^{ik_n x - q_n z} + c_X^{\infty}, \quad X = A, B. \tag{4.6}$$

The third concentration $c_C$ follows from the constraint of Eq. (4.3). In Eq. (4.6), $k_n = 2\pi n/\lambda$ are wave numbers and $q_n$ can be determined from the solutions of the stationary diffusion equation

$$v \partial_z c_X + D \nabla^2 c_X = 0,$$

which yields

$$q_n = \frac{v}{2D} + \sqrt{k_n^2 + \left(\frac{v}{2D}\right)^2}.$$

For all the modes $n \neq 0$, we thus have $q_n \simeq |k_n|$ for small Peclet number $\mathrm{Pe} = \lambda/\ell \ll 1$ with $\ell = 2D/v_n$, which will always be the case for slow growth. The mode $n = 0$ describes the concentration boundary layer which is present at off-eutectic concentrations, and which has a characteristic length scale of $\ell$. To determine the coefficients $X_n$ in the above Fourier series, we assume the eutectic front to be at the $z = 0$ position. Using the Stefan condition in Eq. (4.5) and taking the derivative of $c_X$ with respect to the $z$-coordinate

$$\partial_z c_X|_{z=0} = \sum_{n=-\infty}^{\infty} -q_n X_n e^{ik_n x},$$

integration across one lamella period $\lambda$ of arbitrary partitioning of phases gives

$$q_n X_n \delta_{nm} \lambda = \frac{2}{\ell} \sum_{j=0}^{M-1} \int_{x_j \lambda}^{x_{j+1}\lambda} e^{-ik_m x} \Delta c_X^{\nu_j} \, dx, \tag{4.7}$$

so that the coefficients $X_n, n \in \mathbb{N}$ in the series ansatz, Eq. (4.6) follow

$$X_n = \frac{4}{\ell q_n \lambda k_n} \sum_{j=0}^{M-1} \Delta c_X^{\nu_j} e^{-i k_n \lambda (x_{j+1} + x_j)/2} \sin(k_n \lambda (x_{j+1} - x_j)/2)$$

$$\tag{4.8}$$

Applying symmetry arguments for the sinus and cosinus functions, we can formulate real combinations of these coefficients if we additionally take the negative summation indices into account. We obtain

$$X_n + X_{-n} = \frac{8}{\ell q_n \lambda k_n} \sum_{j=0}^{M-1} \Delta c_X^{\nu_j} \cos(k_n \lambda (x_{j+1} + x_j)/2) \sin(k_n \lambda (x_{j+1} - x_j)/2),$$

$$i(X_n - X_{-n}) = \frac{8}{\ell q_n \lambda k_n} \sum_{j=0}^{M-1} \Delta c_X^{\nu_j} \sin(k_n \lambda (x_{j+1} + x_j)/2) \sin(k_n \lambda (x_{j+1} - x_j)/2).$$

Herewith, Eq. (4.6) reads:

$$c_X = c_X^\infty + X_0 + \sum_{j=0}^{M-1} \sum_{n=1}^\infty \frac{8}{\ell q_n \lambda k_n} \cos(k_n \lambda (x_{j+1} + x_j)/2) \times$$

$$\sin(k_n \lambda (x_{j+1} - x_j)/2) \cos(k_n x) +$$

$$\sum_{j=0}^{M-1} \sum_{n=1}^\infty \frac{8}{\ell q_n \lambda k_n} \sin(k_n \lambda (x_{j+1} + x_j)/2) \sin(k_n \lambda (x_{j+1} - x_j)/2) \sin(k_n x).$$

The general expression for the mean concentration $\langle c_X \rangle_m$ ahead of the $m$-th phase of the phase sequence can be calculated to yield

$$\langle c_X \rangle_m = \frac{1}{(x_{m+1} - x_m)\lambda} \int_{x_m \lambda}^{x_{m+1}\lambda} c_X \, dx$$

$$= c_X^\infty + X_0 + \frac{1}{x_{m+1} - x_m} \sum_{n=1}^\infty \sum_{j=0}^{M-1} \left\{ \frac{16}{\lambda^2 k_n^2 \ell q_n} \Delta c_X^{\nu_j} \right.$$

$$\left. \sin[\pi n(x_{m+1} - x_m)] \times \sin[\pi n(x_{j+1} - x_j)] \cos[\pi n(x_{m+1} + x_m - x_{j+1} - x_j)] \right\}.$$

$$\tag{4.9}$$

For a repetitive appearance of a phase $\nu$ in the phase sequence, the mean concentration of component $X$ ahead of this phase follows by taking the weighted average of all the lamellae of phase $\nu$,

$$\langle c_X \rangle_\nu = \frac{\sum_{m=0}^{M-1} \langle c_X \rangle_m (x_{m+1} - x_m)\delta_{\nu_m \nu}}{\sum_{m=0}^{M-1} (x_{m+1} - x_m)\delta_{\nu_m \nu}} \quad \text{with} \quad \delta_{\nu_m \nu} = \left\{ \begin{array}{ll} 1 & \text{for } \nu = \nu_m \\ 0 & \text{for } \nu \neq \nu_m. \end{array} \right.$$

## 4.2.2. Average front temperature

The average front temperature is now found by taking the average of the Gibbs-Thomson equation along the front, separately for each phase ($\alpha, \beta$ and $\gamma$):

$$\Delta T_\nu = T_E - T_\nu = -m_B^\nu(\langle c_B \rangle_\nu - c_B^E) - m_C^\nu(\langle c_C \rangle_\nu - c_C^E) + \Gamma_\nu \langle \kappa \rangle_\nu, \tag{4.10}$$

for $\nu = \alpha, \beta, \gamma$. Here, $\langle \kappa \rangle_\nu$ is the average curvature of the solid-liquid interface which can be evaluated by exact geometric relations to be

$$\langle \kappa \rangle_\nu = \frac{\sum_{m=0}^{M-1} \langle \kappa \rangle_m (x_{m+1} - x_m)\delta_{\nu_m \nu}}{\sum_{m=0}^{M-1} (x_{m+1} - x_m)\delta_{\nu_m \nu}}$$

and

$$\langle \kappa \rangle_m = \frac{\sin \theta_{\nu_m \nu_{m+1}} + \sin \theta_{\nu_m \nu_{m-1}}}{(x_{m+1} - x_m)\lambda}.$$

Here, $\theta_{\nu_m \nu_{m-1}}$ are the contact angles that are obtained by applying Young's law at the trijunction points. More precisely, $\theta_{\nu_m \nu_{m+1}}$ is the angle, at the triple point (identified by the intersection of the two solid-liquid interfaces and the solid-solid one), between the tangent to the $\nu_m - l$ interface and the horizontal (the $x$ direction). For a triple point with the phases $\nu_m, \nu_{m+1}$ and liquid, the two contact angles $\theta_{\nu_m \nu_{m+1}}, \theta_{\nu_{m+1} \nu_m}$ satisfy the following relations, obtained from Young's law,

$$\frac{\tilde{\sigma}_{\nu_{m+1} l}}{\cos(\theta_{\nu_m \nu_{m+1}})} = \frac{\tilde{\sigma}_{\nu_m l}}{\cos(\theta_{\nu_{m+1} \nu_m})} = \frac{\tilde{\sigma}_{\nu_m \nu_{m+1}}}{\sin(\theta_{\nu_m \nu_{m+1}} + \theta_{\nu_{m+1} \nu_m})}.$$

Note that, in general, $\theta_{\nu_m \nu_{m+1}} \neq \theta_{\nu_{m+1} \nu_m}$.

A short digression is in order to motivate the closure of our system of equations. Although we have not given the explicit expressions, the coefficients $A_0$ and $B_0$ can be simply calculated by using Eq.(4.7) with $n = 0$. However, to carry out this calculation, the width of each lamella has to be given. If these widths are chosen consistent with the lever rule, that is, the cumulated lamellar width of phase $\nu$ corresponds to the nominal volume fraction of phase $\nu$ for the given sample concentration $c_A^\infty$, $c_B^\infty$, and $c_C^\infty$, the use of Eq.(4.7) yields $X_0 = c_X^E - c_X^\infty$ ($X = A, B, C$). However, this result is incorrect: the concentrations of the solids are not equal to the equilibrium concentrations at the eutectic temperature because solidification takes place at a temperature below $T_{\mathbf{E}}$. Therefore, the true volume fractions depend on the solidification conditions. Their determination would require a self-consistent calculation which is exceedingly difficult. Therefore, we will take the same path as Jackson and Hunt in their original paper [44]: we will assume that the volume fractions of the three phases are fixed by the lever rule at the eutectic temperature, but we will treat the amplitudes of the two boundary layers, $A_0$ and $B_0$, as unknowns. As in Ref. [44], one can expect that the difference to the true solution is of order Pe and therefore small for slow solidification.

With this assumption, the equations developed above can now be used in two ways. For *isothermal solidification*, the temperatures of all interfaces must be equal to the externally set temperature, and the three equations $\Delta T_\nu = \Delta T$ for $\nu = \alpha, \beta, \gamma$, can be used to determine the three unknowns $A_0, B_0$ and the velocity $v$ of the solid-liquid front. All of these quantities will be a function of the lamellar spacing $\lambda$. In *directional solidification*, the growth velocity in steady state is fixed and equal to the speed with which the sample is pulled from a hot to a cold region. The third unknown is now the total front undercooling. In the classic Jackson-Hunt theory for binary eutectics, the system of equations is closed by the hypothesis that the average undercoolings of the two phases are equal. This is only an approximation which is quite accurate for eutectics with comparable volume fractions of the two solids, but becomes increasingly inaccurate when the volume fractions are asymmetric [51]. We will use the same approximation for the ternary case here, and set $\Delta T_\alpha = \Delta T_\beta = \Delta T_\gamma = \Delta T$. This then leads to expressions for $\Delta T$ as a function of the growth speed $v$ and the lamellar spacing $\lambda$.

### 4.2.3. Examples

**Binary systems**

As a benchmark for both our calculations and simulations, we consider binary eutectic systems with components $A$ and $B$ and with three phases: $\alpha, \beta$, and liquid. Setting $x_0 = 0, x_1 = \eta_\alpha, x_2 = 1$, and applying Eq. (4.9)

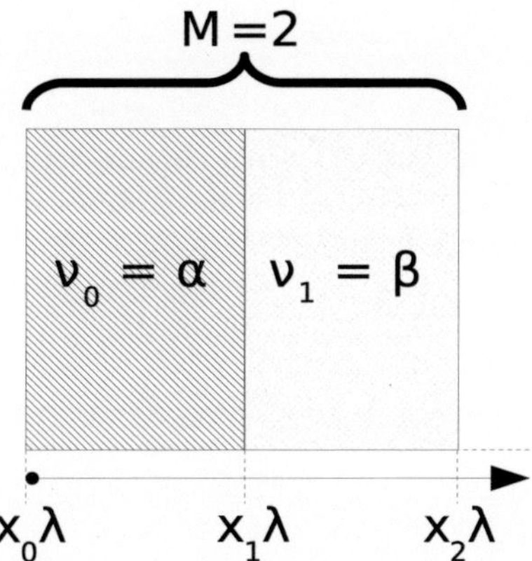



**Figure 4.3.:** Sketch of a lamellar structure in a binary eutectic system with period length $M = 2$. $\nu_i$ denotes a phase in the sequence $(\alpha\beta)$.

gives

$$\langle c_X \rangle_\alpha = c_X^\infty + X_0 + \frac{1}{\eta_\alpha} \sum_{n=1}^{\infty} \left\{ \frac{16}{\lambda^2 k_n^2 \ell q_n} \left( \Delta c_X^\alpha - \Delta c_X^\beta \right) \sin^2(\pi n \eta_\alpha) \right\} \quad (4.11)$$

$$\cong c_X^\infty + X_0 + \frac{2\lambda}{\eta_\alpha \ell} \mathcal{P}(\eta_\alpha) \Delta c_X \qquad \text{and} \qquad (4.12)$$

$$\langle c_X \rangle_\beta = c_X^\infty + X_0 - \frac{2\lambda}{(1 - \eta_\alpha)\ell} \mathcal{P}(1 - \eta_\alpha) \Delta c_X \qquad (4.13)$$

with $k_n = 2\pi n/\lambda, q_n \approx k_n, \lambda/\ell \ll 1, \Delta c_X = \Delta c_X^\alpha - \Delta c_X^\beta$, and the dimensionless function

$$\mathcal{P}(\eta) \;=\; \sum_{n=1}^{\infty} \frac{1}{(\pi n)^3} \sin^2(\pi n\eta) \qquad (4.14)$$

which has the properties $\mathcal{P}(\eta) = \mathcal{P}(1-\eta) = \mathcal{P}(\eta-1)$.

Furthermore, Eq. (4.10) together with $\ell = 2D/v$ leads to

$$\Delta T_\alpha = -m_B^\alpha B_0 - \frac{\lambda v}{\eta_\alpha D}\mathcal{P}(\eta_\alpha)m_B^\alpha \Delta c_B + \Gamma_\alpha \langle \kappa \rangle_\alpha,$$

$$\Delta T_\beta = -m_A^\beta A_0 - \frac{\lambda v}{\eta_\beta D}\mathcal{P}(\eta_\beta)m_A^\beta \Delta c_A + \Gamma_\beta \langle \kappa \rangle_\beta,$$

where $\langle \kappa \rangle_\alpha = 2\sin\theta_{\alpha\beta}/(\eta_\alpha \lambda)$ and $\langle \kappa \rangle_\beta = 2\sin\theta_{\beta\alpha}/(\eta_\beta \lambda)$. In addition, for a binary alloy $B_0 = -A_0$. The unknown $A_0$ and the global front undercooling are determined using the assumption of equal interface undercoolings, $\Delta T_\alpha = \Delta T_\beta$. The result is identical to the one of the Jackson-Hunt analysis.


## Ternary Systems


Next, we study ternary systems with three components $(A, B, C)$ and four phases $(\alpha, \beta, \gamma$ and liquid). We start with the configuration $(\alpha\beta\gamma\alpha\beta\gamma \ldots)$, sketched in Figure 4.4.

We set $x_0 = 0, x_1 = \eta_\alpha, x_2 = \eta_\alpha + \eta_\beta = 1 - \eta_\gamma$ and $x_3 = 1$ and apply Eq. (4.9). This yields

$$\langle c_X \rangle_\alpha \;=\; c_X^\infty + X_0 + \frac{2\lambda}{\eta_\alpha \ell}\left(\mathcal{P}(\eta_\alpha)\Delta c_X^\alpha + \mathcal{Q}(\eta_\alpha, \eta_\beta)\Delta c_X^\beta + \mathcal{Q}(\eta_\alpha, \eta_\gamma)\Delta c_X^\gamma\right)$$

$$\langle c_X \rangle_\beta \;=\; c_X^\infty + X_0 + \frac{2\lambda}{\eta_\beta \ell}\left(\mathcal{Q}(\eta_\beta, \eta_\alpha)\Delta c_X^\alpha + \mathcal{P}(\eta_\beta)\Delta c_X^\beta + \mathcal{Q}(\eta_\beta, \eta_\gamma)\Delta c_X^\gamma\right)$$

$$\langle c_X \rangle_\gamma \;=\; c_X^\infty + X_0 + \frac{2\lambda}{\eta_\gamma \ell}\left(\mathcal{Q}(\eta_\gamma, \eta_\alpha)\Delta c_X^\alpha + \mathcal{Q}(\eta_\gamma, \eta_\beta)\Delta c_X^\beta + \mathcal{P}(\eta_\gamma)\Delta c_X^\gamma\right)$$

$$\cdot \qquad (4.15)$$

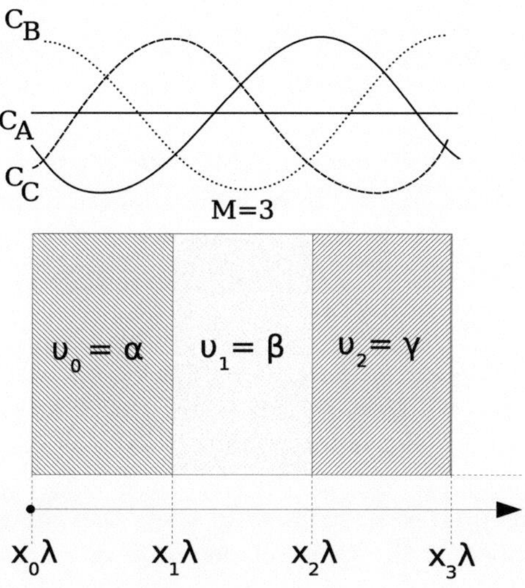


**Figure 4.4.:** Sketch of a ternary stacking order $(\alpha\beta\gamma)$ with period length $M = 3$.

Here, we have used $X = A, B, C$ and $\mathcal{P}$ is the function defined in Eq. (4.14), and

$$\mathcal{Q}(\eta_{\nu_i}, \eta_{\nu_j}) \;=\; \sum_{n=1}^{\infty} \frac{1}{(\pi n)^3} \sin(\pi n \eta_{\nu_i}) \sin(\pi n \eta_{\nu_j}) \cos[\pi n(\eta_{\nu_i} + \eta_{\nu_j})]$$

$\mathcal{P}(\eta_{\nu_i})$ and $\mathcal{Q}(\eta_{\nu_i}, \eta_{\nu_j})$ fulfill the properties $\mathcal{P}(\eta_{\nu_i}) = -\mathcal{Q}(\eta_{\nu_i}, -\eta_{\nu_i})$ and $\mathcal{Q}(\eta_{\nu_i}, \eta_{\nu_j}) = \mathcal{Q}(\eta_{\nu_j}, \eta_{\nu_i})$.

For simplicity, we now consider a completely symmetric ternary eutectic configuration: a completely symmetric ternary phase diagram (that is, any two phases can be exchanged without changing the phase diagram) and equal phase fractions $\eta_\alpha = \eta_\beta = \eta_\gamma = \frac{1}{3}$, which implies $c_X^\infty = c_X^E$. As a consequence, $X_0 = 0$, and Eq. (4.15) simplifies to

$$\langle c_A \rangle_\alpha - c_A^E \;=\; \frac{2\lambda}{\eta_\alpha \ell} \mathcal{P}(\eta_\alpha)(\Delta c_A^\alpha - \Delta c_A^\beta)$$

$$\langle c_B \rangle_\alpha - c_B^E \;=\; \frac{\lambda \mathcal{P}(\eta_\alpha)}{\eta_\alpha \ell}\left(\Delta c_B^\alpha - \Delta c_B^\beta\right)$$

$$\langle c_C \rangle_\alpha - c_C^E \;=\; \frac{\lambda \mathcal{P}(\eta_\alpha)}{\eta_\alpha \ell}\left(\Delta c_C^\alpha - \Delta c_C^\gamma\right),$$

for the three components. Since, in this case, all phases have the same undercooling by symmetry, the front undercooling is simply given by

$$\Delta T = -\frac{2\lambda v}{\eta_\alpha D}\mathcal{P}(\eta_\alpha)m_B^\alpha \Delta c_B + \Gamma_\alpha \langle \kappa \rangle_\alpha$$

where $\langle \kappa \rangle_\alpha = \frac{2}{\eta_\alpha \lambda}(\sin\theta_{\alpha\beta} + \sin\theta_{\alpha\gamma})$. The terms $\Delta c_B^\alpha - \Delta c_B^\beta$ and $\Delta c_C^\alpha - \Delta c_C^\gamma$ are identical. For convenience, we write the preceding equation using the term we already use for the binaries namely $\Delta c_B = \Delta c_B^\alpha - \Delta c_B^\beta$.

Next, we discuss again a ternary eutectic alloy with three components and four phases, but now for the phase cycle $(\alpha\beta\alpha\gamma\alpha\beta\ldots)$. Furthermore, we

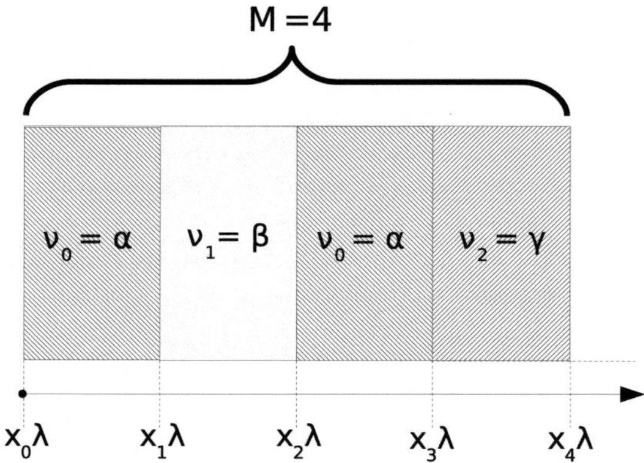


**Figure 4.5.:** Schematic drawing of a ternary eutectic system with a configuration $(\alpha\beta\alpha\gamma\alpha\beta\ldots)$ of periodic length $M = 4$.

suppose that the two lamellae of the $\alpha$ phase have equal width $\lambda\eta_\alpha/2$. The average concentrations $\langle c_X \rangle_m$ are deduced from the general expression in Eq. 4.9 and read

$$\langle c_X \rangle_\alpha \;=\; c_X^\infty + X_0 + \frac{2\lambda}{\frac{\eta_\alpha}{2}\ell}\left(\mathcal{S}(\eta_\alpha,\eta_\beta)\Delta c_X^\alpha + \mathcal{Q}(\tfrac{\eta_\alpha}{2},\eta_\beta)\Delta c_X^\beta + \mathcal{Q}(\tfrac{\eta_\alpha}{2},\eta_\gamma)\Delta c_X^\gamma\right)$$

$$\langle c_X \rangle_\beta \;=\; c_X^\infty + X_0 + \frac{2\lambda}{\eta_\beta \ell}\left(2\mathcal{Q}(\eta_\beta, \tfrac{\eta_\alpha}{2})\Delta c_X^\alpha + \mathcal{P}(\eta_\beta)\Delta c_X^\beta + \mathcal{R}(\eta_\beta, \eta_\gamma)\Delta c_X^\gamma\right)$$

$$\langle c_X \rangle_\gamma \;=\; c_X^\infty + X_0 + \frac{2\lambda}{\eta_\gamma \ell}\left(2\mathcal{Q}(\eta_\gamma, \tfrac{\eta_\alpha}{2})\Delta c_X^\alpha + \mathcal{R}(\eta_\gamma, \eta_\beta)\Delta c_X^\beta + \mathcal{P}(\eta_\gamma)\Delta c_X^\gamma\right),$$

where $X = A, B, C$. Furthermore, we have introduced the short notations

$$\mathcal{R}(\eta_{\nu_i}, \eta_{\nu_j}) \;=\; \sum_{n=1}^{\infty} \frac{1}{(\pi n)^3} \sin(\pi n \eta_{\nu_i}) \sin(\pi n \eta_{\nu_j}) \cos(\pi n)$$

$$\mathcal{S}(\eta_{\nu_i}, \eta_{\nu_j}) \;=\; \sum_{n=1}^{\infty} \frac{1}{(\pi n)^3} \sin^2(\pi n \eta_{\nu_i}/2)\{1 + \cos(\pi n) \cos[\pi n(\eta_{\nu_j} - \eta_{\nu_i})]\}.$$

From the general formulation of the Gibbs-Thomson equation in Eq. (4.10), we determine the undercoolings,

$$\begin{aligned}
\Delta T_\alpha \;=\; & -m_B^\alpha \left(B_0 + \frac{4\lambda}{\eta_\alpha \ell}\left(\mathcal{S}(\eta_\alpha, \eta_\beta)\Delta c_B^\alpha + \mathcal{Q}(\tfrac{\eta_\alpha}{2}, \eta_\beta)\Delta c_B^\beta + \mathcal{Q}(\tfrac{\eta_\alpha}{2}, \eta_\gamma)\Delta c_B^\gamma\right)\right) \\
& + \; -m_C^\alpha \left(C_0 + \frac{4\lambda}{\eta_\alpha \ell}\left(\mathcal{S}(\eta_\alpha, \eta_\beta)\Delta c_C^\alpha + \mathcal{Q}(\tfrac{\eta_\alpha}{2}, \eta_\beta)\Delta c_C^\beta + \mathcal{Q}(\tfrac{\eta_\alpha}{2}, \eta_\gamma)\Delta c_C^\gamma\right)\right) \\
& + \; \Gamma_\alpha \frac{2\left(\sin\theta_{\alpha\beta} + \sin\theta_{\alpha\gamma}\right)}{\eta_\alpha \lambda}
\end{aligned}$$

$$\tag{4.16}$$

$$\begin{aligned}
\Delta T_\beta \;=\; & -m_A^\beta \left(A_0 + \frac{2\lambda}{\eta_\beta \ell}\left(2\mathcal{Q}(\eta_\beta, \tfrac{\eta_\alpha}{2})\Delta c_A^\alpha + \mathcal{P}(\eta_\beta)\Delta c_A^\beta + \mathcal{R}(\eta_\beta, \eta_\gamma)\Delta c_A^\gamma\right)\right) \\
& + \; -m_C^\beta \left(C_0 + \frac{2\lambda}{\eta_\beta \ell}\left(2\mathcal{Q}(\eta_\beta, \tfrac{\eta_\alpha}{2})\Delta c_C^\alpha + \mathcal{P}(\eta_\beta)\Delta c_C^\beta + \mathcal{R}(\eta_\beta, \eta_\gamma)\Delta c_C^\gamma\right)\right) \\
& + \; \Gamma_\beta \frac{2\sin\theta_{\beta\alpha}}{\eta_\beta \lambda}
\end{aligned}$$

$$\begin{aligned}
\Delta T_\gamma \;=\; & -m_A^\gamma \left(A_0 + \frac{2\lambda}{\eta_\gamma \ell}\left(2\mathcal{Q}(\eta_\gamma, \tfrac{\eta_\alpha}{2})\Delta c_A^\alpha + \mathcal{R}(\eta_\gamma, \eta_\beta)\Delta c_A^\beta + \mathcal{P}(\eta_\gamma)\Delta c_A^\gamma\right)\right) \\
& + \; -m_B^\gamma \left(B_0 + \frac{2\lambda}{\eta_\gamma \ell}\left(2\mathcal{Q}(\eta_\gamma, \tfrac{\eta_\alpha}{2})\Delta c_B^\alpha + \mathcal{R}(\eta_\gamma, \eta_\beta)\Delta c_B^\beta + \mathcal{P}(\eta_\gamma)\Delta c_B^\gamma\right)\right) \\
& + \; \Gamma_\gamma \frac{2\sin\theta_{\gamma\alpha}}{\eta_\gamma \lambda}.
\end{aligned}$$

For a symmetric phase diagram (all slopes equal, $m_X^{\nu_i} = m$) one can show using the assumption of equal undercooling of all phases that an expression for the global interface undercooling can be derived as $\Delta T = 1/3(\Delta T_\alpha + \Delta T_\beta + \Delta T_\gamma)$ by elimination of the constants $A_0, B_0$ and $C_0$ using the relation $(A_0 + B_0 + C_0) = 0$.

## 4.2.4. Discussion

A point which merits closer attention is the question which of all the possible steady-state configurations exhibits the lowest undercooling. Whereas the general idea that a eutectic system will always select the state of lowest undercooling is wrong (see Sec. 4.5 below), an information about this point constitutes nevertheless a useful starting point. Whereas the general solution to this problem is non-trivial, in the following we present some partial insights.

Let us, for the sake of discussion, first compute the average total curvature undercooling $\Delta T_\kappa$ of an arbitrary arrangement. Consider a configuration of period M having $M_a$ lamella of the $\alpha$ phase, $M_b$ lamella of the $\beta$ phase, and $M_c$ lamella of the $\gamma$ phase, where the integers $M_a$, $M_b$, and $M_c$ add up to M. In a system where all the solid-liquid and solid-solid surface tensions are identical, the total average curvature undercooling $\Delta T_\kappa^\nu$ of each phase $\nu$ is,

$$\Delta T_\kappa^\alpha = \Gamma_\alpha \frac{2\sin\theta}{\lambda} \frac{M_a}{\eta_\alpha}$$

$$\Delta T_\kappa^\beta = \Gamma_\beta \frac{2\sin\theta}{\lambda} \frac{M_b}{\eta_\beta}$$

$$\Delta T_\kappa^\gamma = \Gamma_\gamma \frac{2\sin\theta}{\lambda} \frac{M_c}{\eta_\gamma}.$$

It is remarkable that the average curvature undercooling is independent of the individual widths of each lamella, but depends only on the total volume fraction and the number of lamellae of the specific phase. Furthermore, it is quite clear from the above examples that the final expression for the global average interface undercooling can always be written in the same form as Eq. (4.1). The second term of this expression (that is, the one proportional to $1/\lambda$) can be computed for the case where all Gibbs-Thomson coefficients and liquidus slopes are equal, and reads

$$
\begin{aligned}
\frac{K_2}{\lambda} &= \frac{\Delta T_\kappa^\alpha + \Delta T_\kappa^\beta + \Delta T_\kappa^\gamma}{3} \\
&= \Gamma \frac{2\sin\theta}{3\lambda}\left(\frac{M_a}{\eta_\alpha} + \frac{M_b}{\eta_\beta} + \frac{M_c}{\eta_\gamma}\right).
\end{aligned}
\tag{4.17}
$$

For the special case of a completely symmetric phase diagram and a sample at the eutectic composition, Eqn.(4.17) yields

$$
\frac{K_2}{\lambda} = \Gamma\frac{2\sin\theta}{\lambda}\left(M_a + M_b + M_c\right),
$$

where we have used the fact that $\eta_\alpha = \eta_\beta = \eta_\gamma = 1/3$. Using, $M_a + M_b + M_c = M$, $\frac{K_2}{\lambda} = \Gamma\frac{2\sin\theta}{(\lambda/M)}$. Thus, we see that the magnitude of this term per unit lamella in an arrangement is the same for all the possible arrangements, irrespective of the individual widths of the lamella and the relative positions of the lamellae in a configuration. Moreover, we see that for a general off-eutectic composition, choosing the number of lamellae in the ratio $\eta_\alpha : \eta_\beta : \eta_\gamma$ renders the average curvature undercoolings of all the three phases equal. This condition is, however, relevant only for the special case of identical solid-solid and solid-liquid surface tensions and equal liquidus slopes of the phases. For the case when the solid-liquid and solid-solid surface tensions are unequal, the curvature undercooling is no longer independent of the arrangement of the lamella in the configuration. Hence, the problem of determining the minimum undercooling configuration is complex and no general expression regarding the number, position and widths of lamellae can be derived.

Another point is worth mentioning. Under the assumption that the volume fractions of the solid phases are fixed by the lever rule, the width of the three lamellae in the $\alpha\beta\gamma$ cycle is uniquely fixed by the alloy concentration. However, for the $\alpha\beta\alpha\gamma$ cycle, and more generally for any cycle with $M > 3$, this is not the case any more because there have to be at least two lamellae of the same phase in the cycle. Whereas the cumulated width of these lamellae is fixed by the global concentration, the width of each individual lamella is not. For example, in the $\alpha\beta\alpha\gamma$ cycle at the eutectic concentration $c_A^\infty = c_B^\infty = c_C^\infty = 1/3$, all the configurations $(\xi, 1/3, 1/3 - \xi, 1/3)$ for $0 < \xi < 1/3$ are admissible, where the notation $(\cdot, \cdot, \ldots)$ is a shorthand for the list of the lamella widths $x_{n+1} - x_n$. The number $\xi$ is an internal degree of freedom that can be freely chosen by the system. With our method, the global front undercooling can be calculated for any value of $\xi$. For the $\alpha\beta\alpha\gamma$ cycle, we found that the configuration with equal widths of the $\alpha$ phases ($\xi = 1/6$) was the one with the minimum average front undercooling. This gives a strong indication that this value is stable, and that perturbations of $\xi$ around this value should decay with time. Hence, the analytic expressions given above for the $\alpha\beta\alpha\gamma$ cycle, which are for $\xi = 1/6$, should be the relevant ones.

## 4.3. Phase-field model

### 4.3.1. Model

A thermodynamically consistent phase-field model is used for the present study [33, 79]. The equations are derived from an entropy functional of the form

$$S\left(e, \boldsymbol{c}, \boldsymbol{\phi}\right) \;=\; \int_\Omega \left( s\left(e, \boldsymbol{c}, \boldsymbol{\phi}\right) - \left( \varepsilon a\left(\boldsymbol{\phi}, \nabla\boldsymbol{\phi}\right) + \frac{1}{\varepsilon} w\left(\boldsymbol{\phi}\right) \right) \right) d\Omega \quad (4.18)$$

where $e$ is the internal energy density, $\boldsymbol{c} = (c_i)_{i=1}^{K}$ is a vector of concentration variables, $K$ being the number of components, and $\boldsymbol{\phi} = (\phi_\alpha)_{\alpha=1}^{N}$ is a vector of phase-field variables, $N$ being the number of phases present in the system. $\boldsymbol{\phi}$ and $\boldsymbol{c}$ fulfill the constraints

$$\sum_{i=1}^{K} c_i = 1 \quad \text{and} \quad \sum_{\alpha=1}^{N} \phi_\alpha = 1, \quad\quad (4.19)$$

so that these vectors always lie in $K - 1$- and $N - 1$-dimensional planes, respectively. Moreover, $\varepsilon$ is the small length scale parameter related to the interface width, $s\left(e, \boldsymbol{c}, \boldsymbol{\phi}\right)$ is the bulk entropy density, $a\left(\boldsymbol{\phi}, \nabla\boldsymbol{\phi}\right)$ is the gradient entropy density and $w\left(\boldsymbol{\phi}\right)$ describes the surface entropy potential of the system for pure capillary-force-driven problems.

We use a multi-obstacle potential for $w(\phi)$ of the form

$$w(\phi) = \begin{cases} \dfrac{16}{\pi^2} \displaystyle\sum_{\substack{\alpha,\beta=1 \\ (\alpha<\beta)}}^{N,N} \sigma_{\alpha\beta}\phi_\alpha\phi_\beta + \displaystyle\sum_{\substack{\alpha,\beta,\gamma=1 \\ (\alpha<\beta<\gamma)}}^{N,N,N} \sigma_{\alpha\beta\gamma}\phi_\alpha\phi_\beta\phi_\gamma, & \text{if } \phi \in \Sigma \\[2em] \infty, & \text{elsewhere} \end{cases} \tag{4.20}$$

where $\Sigma = \{\phi \mid \sum_{\alpha=1}^{N} \phi_\alpha = 1$ and $\phi_\alpha \geq 0\}$, $\sigma_{\alpha\beta}$ is the surface entropy density and $\sigma_{\alpha\beta\gamma}$ is a term added to reduce the presence of unwanted third or higher order phase at a binary interface (see below for details).

The gradient entropy density $a(\phi, \nabla\phi)$ can be written as

$$a(\phi, \nabla\phi) = \sum_{\substack{\alpha,\beta=1 \\ (\alpha<\beta)}}^{N,N} \sigma_{\alpha\beta} \left[a_c(q_{\alpha\beta})\right]^2 |q_{\alpha\beta}|^2,$$

where $q_{\alpha\beta} = (\phi_\alpha \nabla\phi_\beta - \phi_\beta \nabla\phi_\alpha)$ is a vector normal to the $\alpha\beta$ interface. The function $a_c(q_{\alpha\beta})$ describes the form of the anisotropy of the evolving phase boundary. For the present study, we assume isotropic interfaces, and hence $a_c(q_{\alpha\beta}) = 1$. Evolution equations for $c$ and $\phi$ are derived from the entropy functional through conservation laws and phenomenological maximization of entropy, respectively [33, 79]. A linearized temperature field with positive gradient $G$ in the growth direction ($z$ axis) is imposed and moved forward with a velocity $v$,

$$T = T_0 + G(z - vt) \tag{4.21}$$

where $T_0$ is the temperature at $z = 0$ at time $t = 0$. The evolution equations for the phase-field variables read

$$\omega\varepsilon\partial_t\phi_\alpha = \varepsilon\left(\nabla \cdot a_{,\nabla\phi_\alpha}(\phi, \nabla\phi) - a_{,\phi_\alpha}(\phi, \nabla\phi)\right) - \frac{1}{\varepsilon}w_{,\phi_\alpha}(\phi) -$$
$$\frac{f_{,\phi_\alpha}(T, c, \phi)}{T} - \Lambda, \tag{4.22}$$

where $\Lambda$ is the Lagrange multiplier which maintains the constraint of Eq. (4.19) for $\phi$, and the constant $\omega$ is the relaxation time of the phase fields. Furthermore, $a_{,\nabla\phi_\alpha}$, $a_{,\phi_\alpha}$, $w_{,\phi_\alpha}$ and $f_{,\phi_\alpha}$ indicate the derivatives of the respective entropy densities with respect to $\nabla\phi_\alpha$ and $\phi_\alpha$. The function $f(T, c, \phi)$ in Eq. (4.22) describes the free energy density, and is related to the entropy density $s(T, c, \phi)$, through the relation $f(T, c, \phi) = e(T, c, \phi) - Ts(T, c, \phi)$, where $e(T, c, \phi)$ is the internal energy density. The free energy density is given by the summation over all bulk free energy contributions $f_\alpha(T, c)$ of the individual phases in the system. We use an ideal solution model,

$$f(T, c, \phi) = \sum_{i=1}^{K} \left( Tc_i \ln c_i + \sum_{\alpha=1}^{N} c_i L_i^\alpha \frac{(T - T_i^\alpha)}{T_i^\alpha} h_\alpha(\phi) \right), \qquad (4.23)$$

where

$$f_\alpha(T, c) = \sum_{i=1}^{K} \left( Tc_i \ln c_i + c_i L_i^\alpha \frac{(T - T_i^\alpha)}{T_i^\alpha} \right) \qquad (4.24)$$

is the free energy density of the $\alpha$ solid phase, and

$$f_l(T, c) = T \sum_{i=1}^{K} \left( c_i \ln(c_i) \right) \qquad (4.25)$$

is the one of the liquid. The parameters $L_i^\alpha$ and $T_i^\alpha$ denote the latent heats and the melting temperatures of the $i^{th}$ component in the $\alpha$ phase, respectively. We choose the liquid as the reference state, and hence $L_i^l = 0$. The function $h_\alpha(\phi)$ is a weight function which we choose to be of the form $h_\alpha(\phi) = \phi_\alpha^2(3 - 2\phi_\alpha)$. Thus, $f = f_\alpha$ for $\phi_\alpha = 1$. Other interpolation functions involving other components of the $\phi$ vector could also be used, but here we restrict ourselves to this simple choice.

The evolution equations for the concentration fields are derived from Eq. (4.18),

$$\partial_t c_i = -\nabla \cdot \left( M_{i0}(c, \phi) \nabla \frac{1}{T} + \sum_{j=1}^{K} M_{ij}(c, \phi) \nabla \left( \frac{1}{T} \frac{\partial f(T, c, \phi)}{\partial c_j} \right) \right).$$

By a convenient choice of the mobilities $M_{ij}(c, \phi)$, self- and interdiffusion in multicomponent systems (including off-diagonal terms of the diffusion matrix) can be modelled. Here, however, we limit ourselves to a diagonal diffusion matrix with all individual diffusivities being equal, which can be achieved by choosing

$$M_{ij}(c, \phi) = D_i(\phi)c_i \left(\delta_{ij} - c_j\right)$$

$$M_{i0}(c, \phi) = M_{0i}(c, \phi) = - \sum_{\alpha=1}^{N} \sum_{j=1}^{K} M_{ji} h_\alpha(\phi) L_i^\alpha.$$

The terms $M_{i0}(c, \phi) = M_{0i}(c, \phi)$ are the mobilities for the concentration current of the component $i$ due to a temperature gradient. The diffusion coefficient is taken as a linear interpolation between the phases, $D_i(\phi) = \sum_{\alpha=1}^{N} D_i^\alpha \phi_\alpha$, where $D_i^\alpha$ is the non-dimensionalized diffusion coefficient of the $i^{th}$ component in the $\alpha$ phase, using the liquid diffusivity $D^l$ as the reference, where the diffusivities of all the components in the liquid phase are assumed to be equal. In the simulations we assume zero diffusivity in the solid, and take the effective diffusivity to be $D_i(\phi) = D^l \phi_l$. The quantity $d^* = \sigma / (R/v_m)$ is used as the reference length scale in the simulations, where the molar volume $v_m$ is assumed to be independent of the concentration. Here, $\sigma$ is one of the surface entropy density parameters introduced in Eq. (4.20), and the surface entropies of all the phases are assumed to be equal. The reference time scale is chosen to be $t^* = d^{*2}/D^l$. The temperature scale is the eutectic temperature corresponding to the three phase stability regions at the three edges of the concentration simplex and is denoted by $T^*$ while the energy scale is given by $RT^*/v_m$.

### 4.3.2. Relation to sharp interface theory

In order to compare our phase-field simulations to the theory outlined in Sec. 2, we need to relate the parameters of the phase-field model to the quantities needed as input for the theory. For some, this is straightforward. For example, all the parameters of the phase diagram (liquidus slopes, coexistence temperatures etc.) can be deduced from the free energy densities of Eqs. (4.23)–(4.25) in the standard way. For others, the correspondence is less immediate. In the following, we will discuss in some detail two quantities that are crucial for the theory: the surface free energies and the latent heats, both needed to calculate the Gibbs-Thomson coefficients in Eq. (4.4).

The surface free energy $\tilde{\sigma}_{\alpha\beta}$ is defined as the interface excess of the thermodynamic potential density that is equal in two coexisting phases. For alloys, this is not the free energy, but the grand potential. Indeed, the equilibrium between two phases is given by $K$ conditions for $K$ components: $K - 1$ chemical potentials (because of the constraint of Eq. (4.3), only $K - 1$ chemical potentials are independent) as well as $f - \sum_{i=1}^{K-1} \mu_i c_i$, which is the grand potential, have to be equal in both phases. This is the mathematical expression of the common tangent construction for binary alloys and the common tangent plane construction for ternary alloys.

The grand potential excess has several contributions. Since $f = e - Ts$, we need to consider the entropy excess. Both the gradient term in the phase fields and the potential $w(\phi)$ present in the entropy functional give a contribution inside the interface. If, along an $\alpha\beta$ interface, all the other phase fields remain exactly equal to zero, then this contribution can be calculated analytically. However, this is generally not the case: in the interface, the phase fields $\phi_\nu$, $\nu \neq \alpha, \beta$ can be different from zero, which corresponds to an "adsorption" of the other phases. Since the grand potential excess has to be calculated along the equilibrium profile of the fields, the presence of extra phases modifies the value of $\tilde{\sigma}_{\alpha\beta}$. The three-phase terms proportional to $\sigma_{\alpha\beta\gamma}$ have been included in the potential function to reduce (or even eliminate) the additional phases. However, the total removal of these phases requires to choose high values of $\sigma_{\alpha\beta\gamma}$. Such high values ($>10$ times the binary constant $\sigma_{\alpha\beta}$) cause the interface to become steeper near the regions of triple points and lines in 2D and 3D, respectively, which is a natural consequence of the fact that the higher order term affects only the points inside the phase-field simplex where three phases are present. The thinning of the interfaces leads to undesirable lattice pinning, which could only be circumvented by a finer discretization. This, however, would lead to a large increase of the computation times. Therefore, if computations are to remain feasible, we have to accept the presence of additional phases in the interfaces.

Furthermore, there is also a contribution due to the chemical part of the free energy functional. This contribution, identified for the first time in

Ref. [58], arises from the fact that the concentrations inside the interface (which are fixed by the condition of constant chemical potentials) do not, in general, follow the common tangent plane, as illustrated schematically in Figure 4.6.

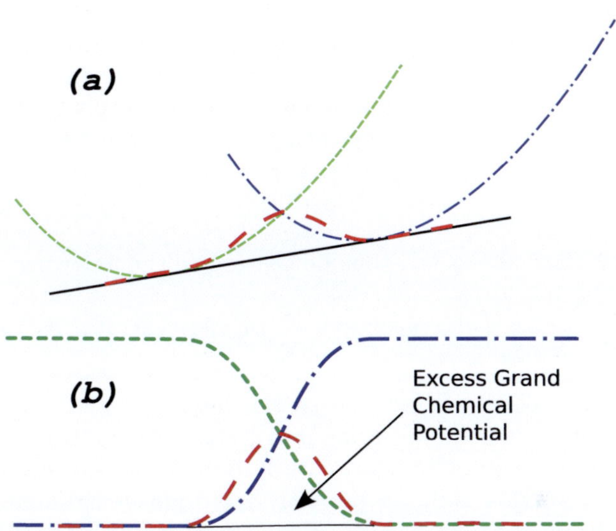


**Figure 4.6.:** Illustration of the existence of an excess interface energy contribution from the chemical free energy. Upper panel: the concentration inside the interfacial region does not necessarily follow the common tangent line. Here, the two convex curves are the free energy densities of the individual phases in contact, the straight line is the common tangent, and the thick non-monotonous line is the concentration along a cut through the interface. Lower panel: the grand chemical potential in the interface differs from the one obtained by a weighted sum of the bulk phase free energies, where the weighting coefficients are the interpolating functions of the order parameters.

Therefore, there is a contribution to the surface free energy which is given by the following expressions. For binary eutectic systems ($N = 3$ phases, $\phi = (\phi_\alpha, \phi_\beta, \phi_l)$; $K = 2$ components $c = (c_A, c_B)$), the vector $c$ is one-dimensional and we define the concentration $(c_A)$ to be the independent field $\boldsymbol{c} = (c_A, 1 - c_A)$. Then, we have

$$\Delta\Psi\left(T, \boldsymbol{c}, \boldsymbol{\phi}\right) = f\left(T, \boldsymbol{c}, \boldsymbol{\phi}\right) - f_l - \mu_A(T)\left(c_A - c_A^l\right),$$

where $\mu_A\left(T\right) = \dfrac{\partial f\left(T, \boldsymbol{c}, \boldsymbol{\phi}\right)}{c_A}$ is the chemical potential of component A. For ternary eutectic systems ($N = 4$ phases, $\boldsymbol{\phi} = (\phi_\alpha, \phi_\beta, \phi_\gamma, \phi_l)$; $K = 3$ components, $\boldsymbol{c} = (c_A, c_B, c_C)$), the vector $\boldsymbol{c}$ is two-dimensional and with the concentrations of $A, B$ as the independent concentration fields, we get $\boldsymbol{c} = (c_A, c_B, 1 - c_A - c_B)$ and the chemical free energy excess becomes,

$$\Delta\Psi\left(T, \boldsymbol{c}, \boldsymbol{\phi}\right) = f\left(T, \boldsymbol{c}, \boldsymbol{\phi}\right) - f_l - \left(\mu_A(T)\right)\left(c_A - c_A^l\right) - \left(\mu_B\left(T\right)\right)\left(c_B - c_B^l\right).$$

The entire surface excess can thus be written as the following

$$\tilde{\sigma}_{\alpha l} \;\; = \;\; \int_x \left(T\varepsilon a\left(\boldsymbol{\phi}, \nabla\boldsymbol{\phi}\right) + \frac{T}{\varepsilon} w\left(\boldsymbol{\phi}\right) + \Delta\Psi\left(T, \boldsymbol{c}, \boldsymbol{\phi}\right)\right) dx$$

where $x$ is the coordinate normal to the interface, and the integral is taken along the equilibrium profile $\boldsymbol{\phi}(x)$, $\boldsymbol{c}(x)$. This integral cannot be calculated analytically. Therefore, we determine the surface free energy numerically. To this end, we perform one-dimensional simulations to determine the equilibrium profiles of concentration and phase fields, and insert the solution into the above formula to calculate $\tilde{\sigma}$. For these simulations, the known bulk values of the concentration fields are used as boundary conditions. To accurately calculate the surface excesses, it is important to include the contribution of the adsorbed phases. For this, the above calculations are performed by letting a small amount of these phases equilibrate at the interface of the major phases. Since the adsorbed phases equilibrate with very different concentrations compared to that of the bulk phases, the domain is chosen large enough such that the chemical potential change of the bulk phases during equilibration is negligible.

Another important quantity which is required as an input in the theoretical expressions is the latent heat of fusion $L_\alpha$ of the $\alpha$ phase. We follow the thermodynamic definition for the latent heat of transformation $L_\alpha$,

$$L_\alpha \;\; = \;\; T_E\left(s_l - s_\alpha\right),$$

$$\text{with} \quad s = -\left(\frac{\partial f\left(T, \boldsymbol{c}, \boldsymbol{\phi}\right)}{\partial T}\right)$$

$$\text{and in particular} \quad s_l = \sum_{i=1}^{K} c_i^l ln\left(c_i^l\right)$$

$$\text{and} \quad s_\alpha = \sum_{i=1}^{K} c_i^\alpha \frac{L_i^\alpha}{T_i^\alpha} + c_i^\alpha ln\left(c_i^\alpha\right),$$

where the concentrations of the phases are taken from the phase diagram at the eutectic temperature.

Finally, let us give a few comments on the interface mobility $\mu_{\text{int}}$ that appears in Eq. (4.4). In early works [13], it was shown that an expression for this mobility in terms of the phase-field parameters can be easily derived in the sharp-interface limit in which the interface thickness tends to zero. Later on, Karma and Rappel [48] proposed the thin-interface limit, in which the interface width remains finite, but much smaller than the mesoscopic diffusion length of the problem. This limit relaxes some of the stringent requirements of the sharp-interface method for the achievement of quantitative simulations. Additionally, this method introduces a correction term to the original expression for the interface mobility, which makes it possible to carry out simulations in the vanishing interface kinetics (infinite interface mobility) regime.

Clearly, such modifications of the interface kinetics are also present in our model, where they arise both from the presence of adsorbed phases in the interface and from the structure of the concentration profile through the interface. Furthermore, it is well known that solute trapping also occurs in phase-field models of the type used here [1]. Since the interface profile can only be evaluated numerically, and since several phase-field and concentration variables need to be taken into account, it is not possible to evaluate quantitatively the contribution of these effects to the interface mobility. However, this lack of knowledge does not decisively impair the present study since we are mainly interested in undercooling versus spacing curves at a fixed interface velocity. At constant velocity, the absolute value of the interface undercooling contains an unknown contribution from the interface kinetics, but the relative comparison between steady states of different spacings remains meaningful. In addition, even though our simulation parameters correspond to higher growth velocities than typical experiments, it will be seen below that the value of the kinetic undercooling in our simulations is small. This indicates once more that our comparisons remain consistent.

## 4.4. Simulation results

In this section, we compare data extracted from phase-field simulations with the theory developed in Sec.4.2, for the case of coupled growth of the solid phases in directional solidification. The simulation setup is sketched in Figure 4.7. Periodic boundary conditions are used in the

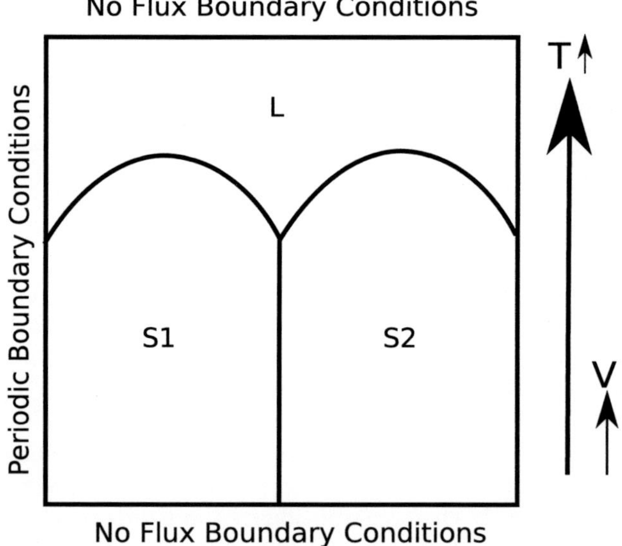


**Figure 4.7.:** Simulation setup for the phase-field simulations of binary and ternary eutectic systems. We impose a temperature gradient $G$ along the $z$ direction and move it with a fixed velocity. The average interface position follows the isotherms at steady state in case of stable lamellar coupled growth.

transverse direction, while no-flux boundary conditions are used in the growth direction. The box width in the transverse direction directly controls the spacing $\lambda$. The box length in the growth direction is chosen several times larger than the diffusion length. The diffusivity in the solid is assumed to be zero. A non-dimensional temperature gradient, G is imposed in the growth direction and moved with a velocity $v$, such that the temperature field is given by Eq. (4.21).

The outline of this section is as follows: first, we will briefly sketch how we extract the front undercooling from the simulation data. Then, this procedure will be validated by comparisons of the results to analytically known solutions as well as to data for binary alloys, for which well-established benchmark results exist. We start the presentation of our results on ternary eutectics by a detailed discussion of the two simplest possible cycles, $\alpha\beta\gamma$ and $\alpha\beta\alpha\gamma$. We compare the data for undercooling as a function of spacing to our analytical predictions and determine the relevant instabilities that limit the range of stable spacings. Finally, we also discuss the behavior of more complicated cycles, for sequences up to length $M = 6$.

## 4.4.1. Data extraction

At steady state, the interface velocity matches the velocity of the isotherms. The undercooling of the solid-liquid interface is extracted at this stage by the following procedure. First, a vertical line of grid points is scanned until the interface is located. Then, the precise position of the interface is determined as the position of the level line $\phi_\alpha = \phi_\beta$ for an $\alpha\beta$-interface (and in an analogous way for all the other interfaces). This is done by calculating the intersection of the phase-field profiles of the corresponding phases, which are extrapolated to subgrid accuracy by polynomial fits. In the presence of adsorbed phases at the interface, the two major phases along the scan line are used for determining the interface point. The major phases are determined from the maximum values that a particular order parameter assumes along the scan line. The temperature at a calculated interface point is then given by Eq. (4.21).

In order to test both our data extraction methods and our calculations of the surface tensions, we have performed the following consistency check.

For an alloy with a symmetric phase diagram at the eutectic concentration, a lamellar front has an equilibrium position when a small temperature gradient ($G = 0.001$) is applied to the system at zero growth speed. Since the concentration in the liquid is uniform for a motionless front, according to the Gibbs-Thomson relation the interface shapes should just be arcs of circles. This was indeed the case in our simulations, and the fit of the interface shapes with circles has allowed us to obtain the interface curvature and the contact angles with very good precision. The extraction of the data is illustrated in Figure 4.8.

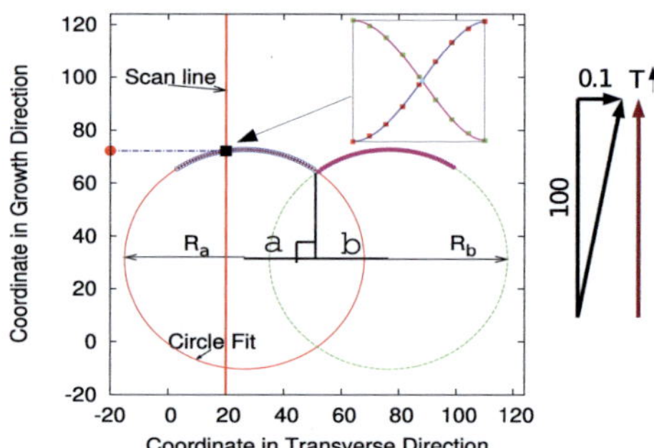


**Figure 4.8.:** Procedure to extract the interface points from the simulation data with sub-grid resolution using higher order interpolation of the phase-field profiles. For the evaluation of the equilibrium properties, the solid-liquid interface points of each lamella are fitted with a circle which is then used to measure the radius of curvature of the particular lamella. We also calculate the triple point angles as the angles between the tangents to the circles at one of the points of intersection.

We fit the radius and the coordinates of the circle centers. Then, the angle at the trijunction point $\theta$ is deduced from geometrical relations, with

$$d = a + b \text{ and } a = \frac{R_a^2 - R_b^2 + d^2}{2d},$$

$$b = d - a$$

$$\theta = \cos^{-1}\left(\frac{a}{R_a}\right) + \cos^{-1}\left(\frac{b}{R_b}\right).$$

The meaning of the lengths $a$ and $b$ is given in Figure 4.8.

## 4.4.2. Validation: Binary Systems

For comparison with the $\Delta T - \lambda$ relationship known from Jackson-Hunt(JH) theory, we create two binary eutectic systems by choosing suitable parameters $L_i^\alpha$ and $T_i^\alpha$ in the free energy density $f(T, \boldsymbol{c}, \boldsymbol{\phi})$. A symmetric binary eutectic system, shown in Figure 4.9a, is created by

$$
L_i^\alpha = \begin{pmatrix} & A & B \\ \alpha & 4.0 & 4.0 \\ \beta & 4.0 & 4.0 \end{pmatrix} \qquad T_i^\alpha = \begin{pmatrix} & A & B \\ \alpha & 1.0 & 0.75980 \\ \beta & 0.75980 & 1.0 \end{pmatrix}.
$$

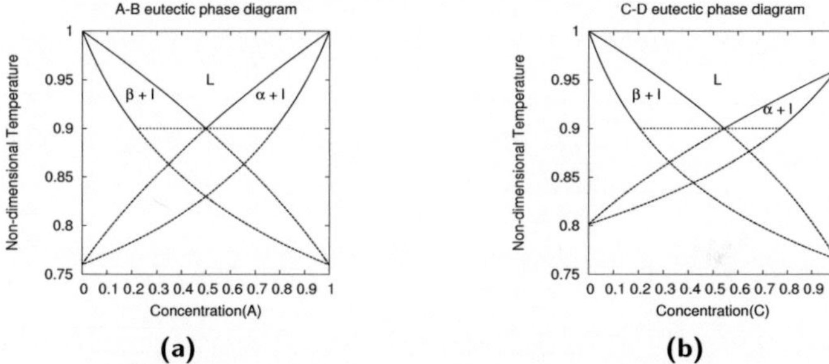


**Figure 4.9.:** Binary eutectic phase diagrams for a model system with stable (solid lines) and metastable (light dashed lines) extensions of the solidus and the liquidus lines, of (a) a symmetric A-B and (b) an unsymmetric C-D system.

To create an asymmetric binary eutectic system, shown in Figure 4.9b, we choose

$$
L_i^\alpha = \begin{pmatrix} & C & D \\ \alpha & 5.0 & 5.0 \\ \beta & 5.0 & 5.0 \end{pmatrix} \qquad T_i^\alpha = \begin{pmatrix} & C & D \\ \alpha & 0.96 & 0.80137 \\ \beta & 0.76567 & 1.0 \end{pmatrix}
$$

The numbers $L_i^\alpha$, $T_i^\alpha$ are chosen such that the widths of each of the (lens-shaped) two-phase coexistence regions remain reasonably broad, and that the approximation of using the values of concentration difference between the solidus and liquidus $\left(\Delta c_\nu^l\right)$ at the eutectic temperature for the theoretical expressions holds for a good range of undercoolings. This implies that the value of the $L_i^\alpha$ should not be too small. Conversely, a too high value is also not desirable since for large values of $L_i^\alpha$ the chemical contribution to the surface free energy becomes large, which leads to very steep and narrow interface profiles.

**Table 4.1.:** Parameters for the sharp-interface theory, with proper calculation of the surface tension in the phase-field simulations for (a) a symmetric binary eutectic system with components A and B and (b) for an unsymmetric binary eutectic system with components C and D

(a)

| | |
|---|---|
| $\widetilde{\sigma}_{\alpha l}$ | 1.01146 |
| $\widetilde{\sigma}_{\beta l}$ | 1.01146 |
| $\widetilde{\sigma}_{\alpha\beta}$ | 1.23718 |
| $\theta_{\alpha\beta}$ | 37.70 |
| $\theta_{\beta\alpha}$ | 37.70 |
| $L^\alpha$ | 4.0 |
| $L^\beta$ | 4.0 |
| $m_B^\alpha = m_A^\beta$ | -0.206975 |

(b)

| | |
|---|---|
| $\widetilde{\sigma}_{\alpha l}$ | 0.97272 |
| $\widetilde{\sigma}_{\beta l}$ | 1.07235 |
| $\widetilde{\sigma}_{\alpha\beta}$ | 1.24836 |
| $\theta_{\alpha\beta}$ | 33.903 |
| $\theta_{\beta\alpha}$ | 41.161 |
| $L^\alpha$ | 4.686 |
| $L^\beta$ | 4.711 |
| $m_D^\alpha$ | -0.13161 |
| $m_C^\beta$ | -0.22138 |

We perform simulations at two different velocities $V = 0.01$ and $V = 0.02$, with a mesh size $\Delta x = 1.0$ and the parameter set $\varepsilon = 4.0, D_A^l = D_B^l = D_C^l = D_D^l = 1.0, \sigma_{\alpha\beta} = \sigma_{\alpha l} = \sigma_{\beta l} = 1.0, \sigma_{\alpha\beta\gamma} = 10.0$. To give an idea of the order of magnitude of the corresponding dimensional quantities, we remark that if we assume the melting temperatures to be around 1700K and the other values to correspond to the Ni-Cu system used in the study of Warren $et\ al.$ [128], the length scale $d^*$ for the case of the binary eutectic system turns out to be around 0.2 nm and the time scale 0.04 ns.

The corresponding parameters for the sharp-interface theory are given in Table 4.1. The comparisons between our numerical results and the analytic theory are shown in Figs. 4.10a and 4.10b.

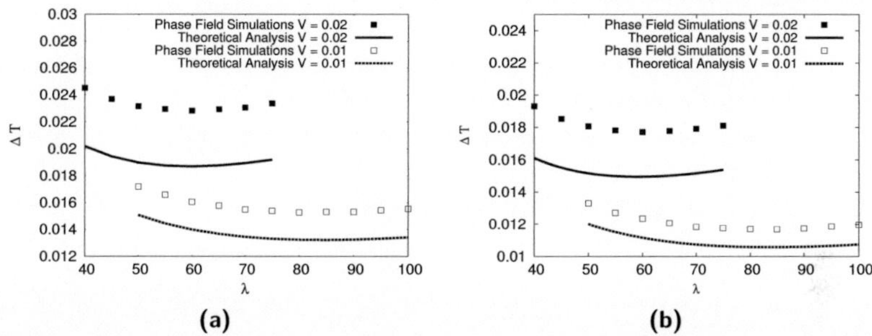


**Figure 4.10.:** Comparison of $\Delta T - \lambda$ relations resulting from the theoretical analysis and from the phase-field simulations at two different velocities for systems; (a) symmetric binary eutectic system (A-B) and (b) unsymmetric binary eutectic system (C-D).

Consistent differences can be observed in the undercooling values between our data and the predictions from JH theory for both systems. The difference in undercoolings is smaller at lower velocities, which hints at the presence of interface kinetics. We find indeed that when we change the relaxation constant in the phase-field evolution equation by about 50 %, the difference between the predicted and measured undercoolings is removed for the case of the considered symmetric binary phase diagram. This clearly shows that the interface kinetics is not negligible. It seems difficult, however, to obtain a precise numerical value for its magnitude in the framework of the present model.

The spacing at minimum undercooling, however, is reproduced to a good degree of accuracy (error of 5 %), while the minimum undercooling has a maximum error of 10 %. It should also be noted that the JH theory only is an approximation for the true front undercooling. Results obtained both with boundary integral [51] and quantitative phase-field methods [32] have shown that, whereas the prediction for the minimum undercooling spacing is excellent, errors of 10 % for the value of the undercooling itself are typical. If the JH curve is drawn without taking into account the additional chemical contributions to the surface tension, a completely different result is obtained, with minimum undercooling spacings that are largely different from the simulated ones. We can therefore conclude that we have captured the principal corrections.

In addition, we have performed equilibrium measurements of the angles at the trijunction point and of the radius of curvature of the lamellae as described in the preceding sub-section (4.4.1) for the symmetric eutectic system. The contact angles differ from the ones predicted by Young's equilibrium conditions only by a value of 0.2 degrees. The theoretical (from the Gibbs-Thomson equation) and measured undercoolings differ in the third decimal, with an error of 0.1 %.

### 4.4.3. Ternary Systems: Parameter set

$$
L_i^\alpha = \begin{pmatrix}
 & A & B & C \\
\alpha & 1.46964038 & 1.0 & 1.0 \\
\beta & 1.0 & 1.46964038 & 1.0 \\
\gamma & 1.0 & 1.0 & 1.46964038
\end{pmatrix}
$$

$$
T_i^\alpha = \begin{pmatrix}
 & A & B & C \\
\alpha & 1.5 & 0.5 & 0.5 \\
\beta & 0.5 & 1.5 & 0.5 \\
\gamma & 0.5 & 1.0 & 1.5
\end{pmatrix}.
$$

We use a symmetric ternary phase diagram. The following matrices list the parameters $L_i^\alpha$, $T_i^\alpha$ in the free energy $f(T, c, \phi)$ that were used to create a symmetric ternary eutectic system, shown in Figure 4.1. We perform simulations with the parameter set $\varepsilon = 8.0, \Delta x = 1.0, D_A^l = D_B^l = D_C^l = 1.0, \sigma_{\alpha\gamma} = \sigma_{\beta\gamma} = \sigma_{\gamma\beta} = \sigma_{\alpha l} = \sigma_{\beta l} = \sigma_{\gamma l} = 1.0, \sigma_{\alpha\beta l} = \sigma_{\alpha\beta\gamma} = \sigma_{\alpha\gamma l} = \sigma_{\beta\gamma l} = 10.0$ and compare with the theoretical expressions using the input parameters listed in Table 4.2.

**Table 4.2.:** Input parameters for the theoretical relations for the ternary eutectic system

| | |
|---|---|
| $\widetilde{\sigma}_{\alpha l} = \widetilde{\sigma}_{\beta l} = \widetilde{\sigma}_{\gamma l}$ | 1.194035 |
| $\widetilde{\sigma}_{\alpha\beta} = \widetilde{\sigma}_{\alpha\gamma} = \widetilde{\sigma}_{\beta\gamma}$ | 1.430923 |
| $\theta_{\alpha\beta} = \theta_{\beta\alpha} = \theta_{\gamma\alpha} = \theta_{\alpha\gamma}$ | 36.81 |
| $L^{\alpha} = L^{\beta} = L^{\gamma}$ | 1.33 |
| $m_B^{\alpha} = m_C^{\alpha}$ | -0.91 |
| $m_A^{\beta} = m_C^{\beta}$ | -0.91 |
| $m_A^{\gamma} = m_B^{\gamma}$ | -0.91 |

## 4.4.4. Simple cycles: steady states and oscillatory instability

We first perform simulations to isolate the regime of stable lamellar growth for the configuration $\alpha\beta\gamma$. For this regime, we measure the average interface undercooling and compare it to our theoretical predictions. The results are shown in Figure 4.11.

The agreement in the undercoolings is much better than for the binary eutectic systems, with a smaller dependence of undercoolings on the velocities. Consequently, both the spacing at minimum undercooling (error 4 % for V=0.005 and  6 % for V=0.01) and the minimum undercooling (error of 1-2 %), match very well with the theoretical relationships, as shown in Figure 4.11. The equilibrium angles at the triple point also agree with the ones predicted from Young's law to within an error of 0.3 degrees, while the radius of curvature matches that from the Gibbs-Thomson relationship with negligible error (<0.5 %).

It should be noted that the steady lamellae remain straight, contrary to the results of Ref. [40], where a spontaneous tilt of the lamellae with respect to the direction of the temperature gradient was reported. This difference is due to the different phase diagrams: we are using a completely symmetric phase diagram and equal surface tensions for all solid-liquid interfaces, whereas [40] uses the thermophysical data of a real alloy.

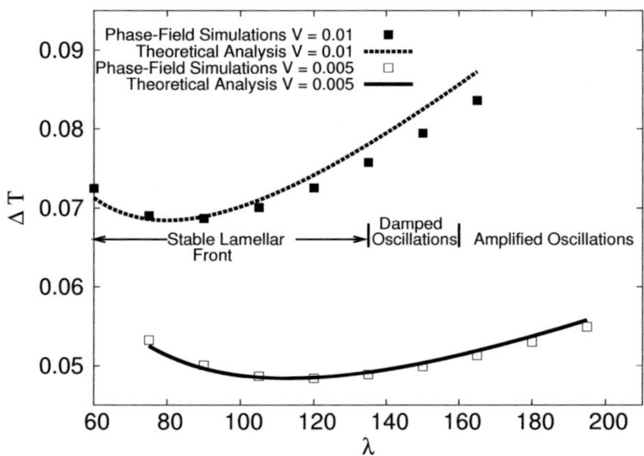


**Figure 4.11.:** Comparison between theoretical analysis and phase-field simulations at two different velocities for the arrangement $(\alpha\beta\gamma)$ of ternary eutectic solids at $V = 0.005$ and $V = 0.01$. The demarcation shows the regions of stable lamellar growth and the critical spacing beyond which we observe amplified oscillatory behavior. There is a small region named "Damped Oscillations", which is a region where oscillations occur but die down slowly with time.

Next, we are interested in the stability range of three-phase coupled growth. From general arguments, we expect a long-wavelength lamella elimination instability (Eckhaus-type instability) to occur for low spacings, as in binary eutectics [2]. Here, we will focus on oscillatory instabilities that occur for large spacings. It is useful to first recall a few facts known about binary eutectics, where all the instability modes have been classified [34, 51]. Lamellar arrays in binary eutectics are characterized (in the absence of crystalline anisotropy) by the presence of two mirror symmetry planes that run in the center of each type of lamellae, as sketched in Figure 4.12(a). Instabilities can break certain of these symmetries while other symmetry elements remain intact [24]. In binary eutectics, the oscillatory 1-$\lambda$-O mode is characterized by an in-phase oscillation of the thickness of all $\alpha$ (and $\beta$) lamellae; both mirror symmetry planes remain in the oscillatory pattern. In contrast, in the 2-$\lambda$-O mode, one type of lamellae start to oscillate laterally, whereas the mirror plane in the other type of lamellae survives; this leads to a spatial period doubling. Finally, in the tilted pattern both mirror planes are lost.

It is therefore important to survey the possible symmetry elements in the ternary case. At first glance, there seems to be no symmetry plane in the pattern. However, for our specific choice of phase diagram, new symmetry elements not present in a generic phase diagram exist: mirror symmetry planes combined with the exchange of two phases. Consider for example the $\beta$ phase in the center of Figure 4.12(b): if the system is reflected at its center, and then the $\alpha$ and $\gamma$ phases are exchanged, we recover the original pattern. At the eutectic concentration, there are three such symmetry planes running in the center of each lamella, and three additional ones running along the three solid-solid interfaces. Off the eutectic point, two of these planes survive if any two of the three phases have equal volume fractions. Guided by these considerations, we can conjecture that there are two obvious possible instability modes, sketched in Figure 4.13.

In the first, called mode 1 in the following, two symmetry planes survive: the width of one lamella oscillates, whereas the two other phases form a "composite lamella" that oscillates in opposition of phase; the interface in the center of this composite lamella does not oscillate at all and constitutes one of the symmetry planes. In the second (mode 2), the lateral position of one of the lamellae oscillates with time, whereas the other two phases oscillate in opposition of phase to form a "composite lamella" that oscillates

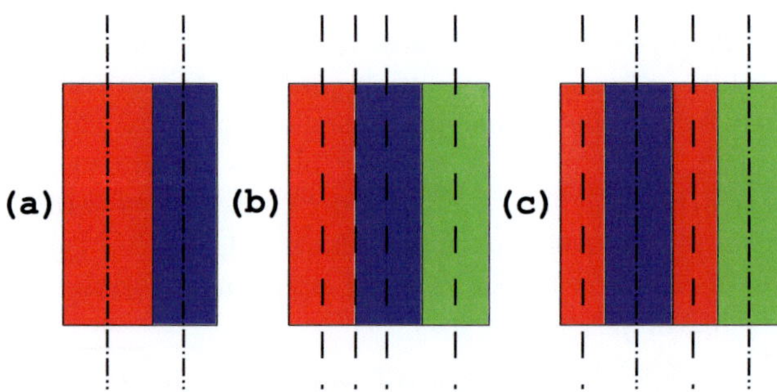

**Figure 4.12.:** In a periodic arrangement of lamellae, we can identify certain lines of symmetry, as shown in (a) for a binary eutectic. Similarly, for the case of the two simplest configurations, (b) $\alpha\beta\gamma$ and (c) $\alpha\beta\alpha\gamma$ in a symmetric ternary eutectic system, such planes of symmetry exist. While in the case of a binary eutectic, the lines are mirror symmetry axes (shown by dash-dotted lines), in the special case of a symmetric ternary phase diagram, one can also identify quasi-mirror lines (dashed lines) where we retrieve the original configuration after a spatial reflection and an exchange of two phases. Only quasi-mirror lines exist in the $\alpha\beta\gamma$ arrangement, which are shown in (b), while both true- and quasi-mirror planes exist in the $\alpha\beta\alpha\gamma$ arrangement as shown in (c).

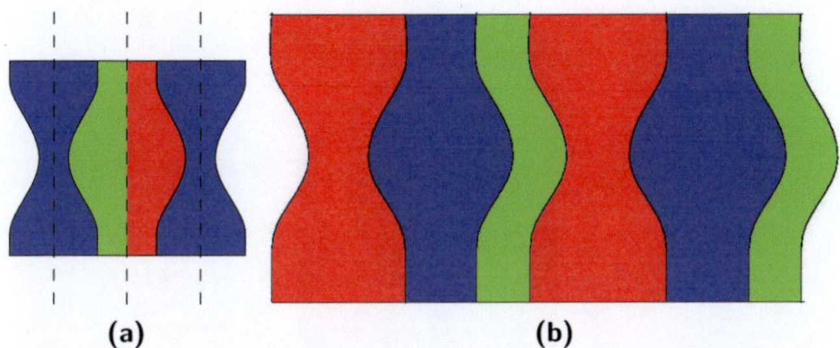


**Figure 4.13.:** Guided by the symmetry axes in the $\alpha\beta\gamma$ arrangement, one can expect two possible oscillatory modes at off-eutectic concentrations along the eutectic groove. The oscillations in (a), which keep all the quasi-mirror planes intact, are expected to occur at concentrations towards the apex of the simplex along the eutectic groove. Another possibility, shown in (b) exists in which no symmetry plane remains, which is expected to occur at a concentration towards the binary edge of the simplex.

laterally but keeps an almost constant width. There is no symmetry plane left in this mode.

The stability range of the coupled growth regime of the lamellar arrangement is indicated in Figure 4.11. Steady lamellar growth is stable from below the minimum undercooling spacing up to a point where an oscillatory instability occurs. In the region marked "damped oscillations", oscillatory motion of the interfaces was noticed, but died out with time. Above a threshold in spacing, oscillations are amplified. We monitored the modes that emerged, and found indeed good examples for the two theoretically expected patterns, shown in Figure 4.14.

Mode 1 is favored for off-eutectic concentrations in which one of the lamellae is wider than the two others, such as $c = (0.32, 0.32, 0.36)$. Indeed, in that case the (unstable) steady-state pattern exhibits the same symmetry planes as the oscillatory pattern. This mode can also appear when one lamella is *smaller* than the two others, see Figure 4.14c. We detect mode 2 at the eutectic concentration, see Figure 4.15b. However, a "mixed mode" can also occur, in which no symmetry plane survives, but the three

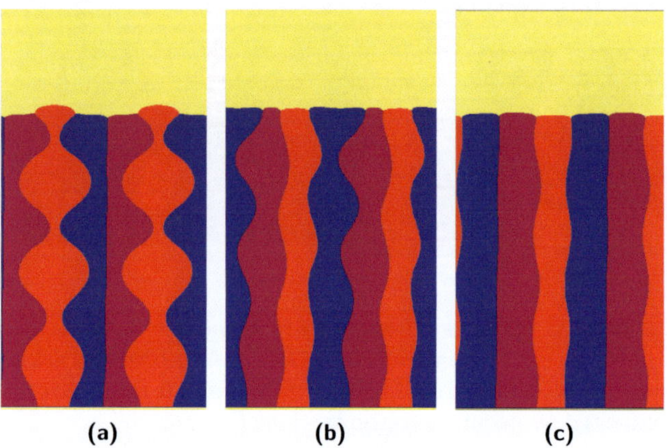

**Figure 4.14.:** Oscillatory modes in simulations for the $\alpha\beta\gamma$ configuration at the off-eutectic concentrations $c = (0.32, 0.32, 0.36))$ in (a), and $c = (0.34, 0.34, 0.32)$ in (b) and (c). The spacings are $\lambda = 170$ in (a) and (c) and $\lambda = 165$ in (b).

trijunctions oscillate laterally with phase differences that depend on the concentration and possibly on the spacing, see Figs. 4.14b and 4.15c.

Let us now turn to the $\alpha\beta\alpha\gamma$ cycle. We perform simulations for two different velocities $V = 0.01$ and $V = 0.005$. The comparison of the measurements with the theoretical analysis for steady-state growth is shown in Figure 4.16. For the purpose of analysis, predictions from the theory for both arrangements ($\alpha\beta\gamma$ and $\alpha\beta\alpha\gamma$) are also shown. Here again, the minimum undercooling spacings match those of the theory to a good degree of accuracy (error 5%, V=0.005). However, the undercooling is lower than the one predicted by JH-theory, with a discrepancy of 4% for the case of $V = 0.005$, Figure 4.16a. For $V = 0.01$, Figure 4.16b, simulations were not possible for a sufficient range of $\lambda$ to determine the minimum undercooling, because the width of the narrowest lamellae became comparable to the interface width $\varepsilon \simeq 8.0$ before the minimum was reached. However, the general trend of the data follows the predictions of the theory for both velocities. This was also the case for simulations carried

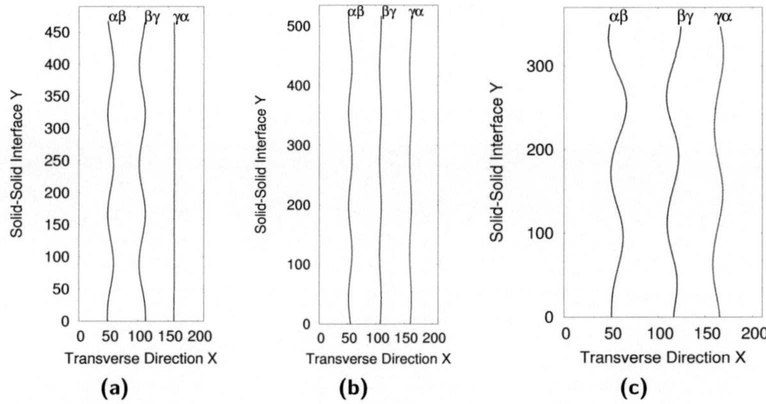


**Figure 4.15.:** The plot shows the trace of the triple points for the $\alpha\beta\gamma$ arrangement. The growth direction is compressed in these plots with respect to the transverse direction in order to better visualize the modes. We get multiple modes at the eutectic concentration for the same spacing $\lambda = 159$, shown in (a) and (b). In (a) we get back mode 1 while (b) matches well to our predicted mode 2. A mixed mode (c) is obtained at an off-eutectic concentration $c = (0.34, 0.34, 0.32)$, at a spacing $\lambda = 165$, which is a combination of oscillations in both the width and lateral spacing.

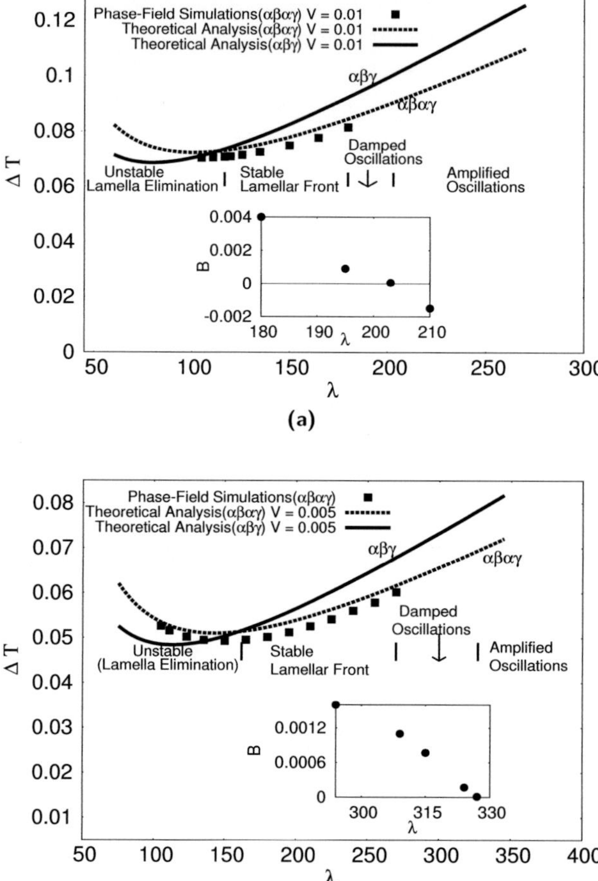



**Figure 4.16.:** Theoretical analysis and phase-field simulations: Comparison between the arrangements $(\alpha\beta\gamma)$ and $(\alpha\beta\alpha\gamma)$ for two different velocities (a) V = 0.005 and (b) V = 0.01. Plots convey information on the stability ranges, and the onset of oscillatory behavior of the 1-$\lambda$-O type.

out at an off-eutectic concentration $\boldsymbol{c} = (0.32, 0.34, 0.34)$ at a velocity of $V = 0.005$, for the same configuration $\alpha\beta\alpha\gamma$.

Concerning the oscillatory instabilities at large spacings, it is useful to consider again the symmetry elements. For this cycle, there are two real symmetry planes in the steady-state pattern that run through the centers of the $\beta$ and $\gamma$ lamellae. Note that these symmetries would exist even for unsymmetric phase diagrams and unequal surface tensions. Therefore, by analogy with binary eutectics, one may expect oscillatory modes that simply generalize the 1-$\lambda$-O and 2-$\lambda$-O modes of binary eutectics, see Figure 4.17a. Indeed, for our simulations at the eutectic concentration, we retrieve the 1-$\lambda$-O type oscillation, figure 4.18a as in our hypothesis (figure 4.17a).

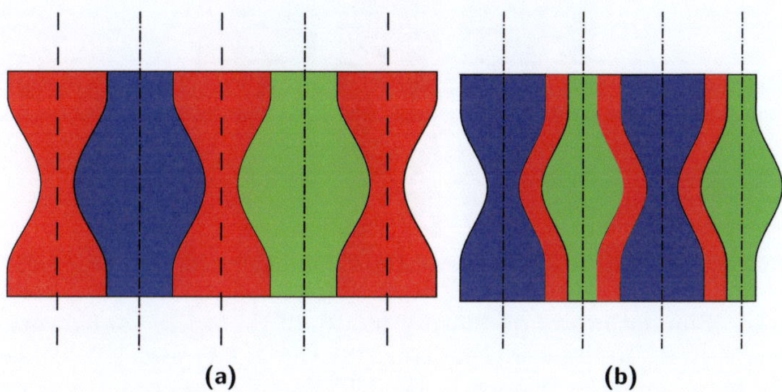

**Figure 4.17.:** Predictions of oscillatory modes for the $\alpha\beta\alpha\gamma$ arrangement, reminiscent of the 1-$\lambda$-O mode (a) and 2-$\lambda$-O mode (b) in binary eutectics.

This oscillatory instability can be quantitatively monitored by following the lateral positions of the solid-solid interfaces with time. More specifically, we extract the width of the $\beta$ phase as a function of the growth distance $z$. This is then fitted with a damped sinusoidal wave of the type $A_0 + A\exp(-Bz)\cos((2\pi z/L) + D)$. The damping coefficient $B$ is obtained from a curve fit and plotted as a function of the spacing $\lambda$. The onset of the instability is characterized by the change in sign of the damping coefficient.

**Figure 4.18.:** Simulations of oscillatory modes of the $\alpha\beta\alpha\gamma$ configuration. The modes in (a) and (b) show resemblance to the 1-$\lambda$-O and 2-$\lambda$-O oscillatory modes of binary eutectics, respectively. Additionally, other modes (c) can also be observed, depending on the initial conditions. While we observe (a) at the eutectic concentration, (b) and (c) are modes at off-eutectic concentrations $c = (0.32, 0.34, 0.34)$. The spacings are (a) $\lambda = 201$, (b) $\lambda = 174$ (c) $\lambda = 210$.

For the off-eutectic concentration we get two modes (figure 4.18). While (figure 4.18b) corresponds well to our hypothesis to the 2-$\lambda$-O type oscillation (figure 4.17b), we also observe another mode as shown in figure 4.18c, which combines elements of the two modes: both the width and the lateral position of the $\alpha$ lamellae oscillate.

## 4.4.5. Lamella elimination instability

For the $\alpha\beta\alpha\gamma$ cycle, there is also a new instability, which occurs for low spacings. We find that all spacings below the minimum undercooling spacing, as well as some spacings above it, are unstable with respect to lamella elimination: the system evolves to the $\alpha\beta\gamma$ arrangement by eliminating one of the $\alpha$ lamellae, both at eutectic and off-eutectic concentrations. The points plotted to the left of the minimum in Figure 4.16b are actually unstable steady states that can be reached only when the simulation is started with strictly symmetric initial conditions and the correct volume fractions of the solid phases.

This instability can actually be well understood using our theoretical expressions. As already mentioned before, when we consider the cycle $\alpha\beta\alpha\gamma$ at the eutectic concentration with a lamella width configuration $(\xi, 1/3, 1/3 - \xi, 1/3)$, the global average front undercooling attains a minimum for the symmetric pattern $\xi = 1/6$. However, the global front undercooling is not the most relevant information for assessing the front stability. More interesting is the undercooling of an *individual* lamella, because this can give information about its evolution. More precisely, consider the undercooling of one of the $\alpha$ lamellae as a function of $\xi$. If the undercooling increases when the lamella gets thinner, then the lamella will fall further behind the front and will eventually be eliminated. In contrast, if the undercooling decreases when the lamella gets thinner, then the lamella will grow ahead of the main front and get larger. A similar argument has been used by Jackson and Hunt for their explanation of the long-wavelength elimination instability [44]. It should be pointed out that the new instability found here is not a long-wavelength instability, since it can occur even when only one unit cell of the cycle is simulated.

Following the above arguments, we have calculated the growth temperature of the first $\alpha$ lamella as a function of $\xi$ using the general expressions in

Eq. 4.16. In Figure 4.19, we plot the variation of $\partial\Delta T/\partial\xi$ at $\xi = 1/6$, as a function of $\lambda$. The point at which $\partial\Delta T/\partial\xi$ becomes positive then indicates the transition to a stable $\alpha\beta\alpha\gamma$ cycle. This criterion is in good agreement with our simulation results. This argument can also be generalized to more complicated cycles (see below).

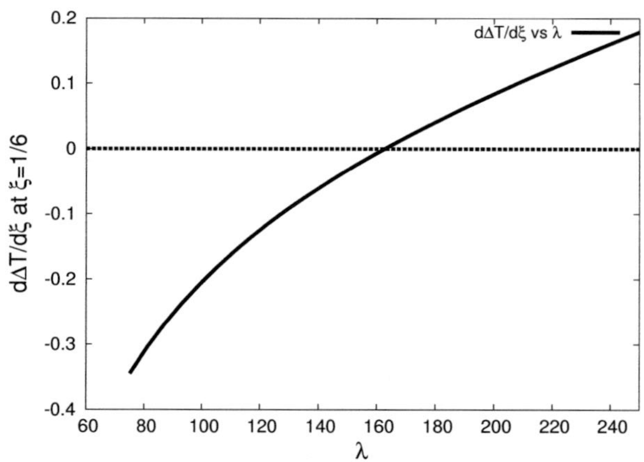


**Figure 4.19.:** Plot of $\partial\Delta T/\partial\xi$, taken at $\xi = 1/6$ versus $\lambda$ for the $\alpha_1\beta\alpha_2\gamma$ cycle, where $\Delta T$ is the undercooling of the $\alpha_1$ lamella and $\xi$ its width (relative to $\lambda$), calculated by our analytical expressions in the volume fraction configuration. $(\xi, 1/3, 1/3 - \xi, 1/3)$ at V=0.01. The cycle is predicted to be unstable to lamella elimination if $\partial\Delta T/\partial\xi < 0$. The $\lambda$ at which $\partial\Delta T/\partial\xi$ changes sign is the critical point beyond which the $\alpha\beta\alpha\gamma$ arrangement is stable with respect to a change to the sequence $\alpha\beta\gamma$ through a lamella elimination.

## 4.4.6. Longer cycles

Let us now discuss a few more complicated cycles. The simple cycles we have simulated until now were such that during stable coupled growth the widths of all the lamellae corresponding to a particular phase were the same. This changes starting from period $M = 5$, where the configuration $\alpha\beta\alpha\beta\gamma$ is the only possibility (up to permutations). If we consider this cycle at the eutectic concentration and note the configuration of lamella widths as $(\xi, 1/3 - \xi, 1/3 - \xi, \xi, 1/3)$ and compute the average front undercooling by our theoretical expressions, we find that the minimum occurs for $\xi$ close to 0.12. In addition, for this configuration, the undercooling of any asymmetric configuration (permutation of widths of lamellae) is higher than the one considered above. If we rewrite symbolically this configuration as $\alpha_1, \beta_2, \alpha_2, \beta_1, \gamma$, it is easy to see that this configuration has two symmetry axes of the same kind as discussed in the preceding subsection: mirror reflection and exchange of the phases $\alpha$ and $\beta$. One of them runs along the interface between $\beta_2$ and $\alpha_2$, and the other one in the center of the $\gamma$ lamella.

Not surprisingly, our simulation results confirm the importance of this symmetry. The volume fractions in steady-state growth are close to those that give the minimum average front undercooling, see Figs. 4.20b and 4.20c. Additionally, we observe oscillations in the width of the largest $\gamma$

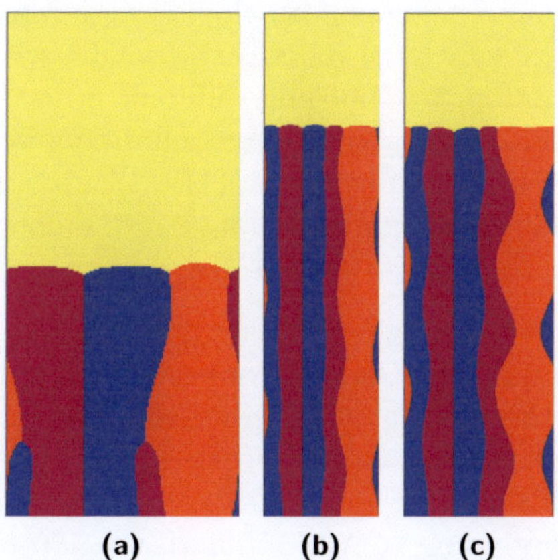


**Figure 4.20.:** Simulations at spacings $\lambda = 135$ in (a), $\lambda = 150$ in (b) and $\lambda = 180$ in (c), starting from an initial configuration of $\alpha\beta\alpha\beta\gamma$. There is no spacing for which the $\alpha\beta\alpha\beta\gamma$ develops into a stable lamellar growth front. Smaller spacings switch to the $\alpha\beta\gamma$ arrangement while the larger spacings exhibit oscillatory instabilities in both the width and the lateral positions of the lamellae.

phase and oscillations in the widths and the lateral position of the smaller lamellae of the $\alpha$ and $\beta$ phases, while the interface between the larger $\alpha$ and $\beta$ phase remains straight, such that the combination of all the $\alpha$ and $\beta$ lamellae oscillates in width as one "composite lamella". Thus, the symmetry elements of the underlying steady state are preserved in the oscillatory state.

For smaller spacings, this configuration is unstable, and the sequence changes to the $\beta\alpha\gamma$ arrangement as shown in Figure 4.20a by two successive lamella eliminations. It is noteworthy that we did not find any unstable sequence which switches to the $\alpha\beta\alpha\gamma$ arrangement, which again can be understood from the presence of the symmetry. Indeed, a symmetrical evolution would result in a change to a configuration $\alpha_1\beta_1\gamma$ or $\beta_2\alpha_2\gamma$, but precludes the change to a configuration of period length $M = 4$.

Going on to cycles with period $M = 6$, the first arrangement we consider is $\alpha_1\beta_1\alpha_2\beta_2\alpha_1\gamma$, where we name the lamellae for eventual discussion and ease in description according to the symmetries. Indeed, this arrangement has two exact mirror symmetry planes in the center of the $\alpha_2$ and the $\gamma$ phases. We find that, if we calculate the average interface undercooling curves by varying the widths of individual lamella with the constraint of constant volume fraction, by choosing different $\xi$, in the width configuration $(\xi, 1/6, 1/3 - 2\xi, 1/6, \xi, 1/3)$, the average undercooling at the growth interface is minimal for the configuration $(1/9, 1/6, 1/9, 1/6, 1/9, 1/3)$. This arrangement has the highest undercooling curve among the arrangements we have considered, shown in Figure 4.21a.

It also has a very narrow range of stability, and we could isolate only one spacing which exhibits stable growth for $\lambda = 240$, Figure 4.22d. Unstable arrangements near the minimum undercooling spacing evolve to the $\alpha\beta\gamma$ arrangement, Figure 4.22a, while for other unstable configurations we obtain the arrangements in Figure 4.22b and Figure 4.22c as the stable growth forms corresponding to $\lambda = 150$ and $\lambda = 180$ respectively.

Apart from the (trivial) period-doubled arrangement $\alpha\beta\gamma\alpha\beta\gamma$, another possibility for $M = 6$ is $\alpha\beta\gamma\alpha\gamma\beta$ with a volume fraction configuration $(1/6, 1/6, 1/6, 1/6, 1/6, 1/6)$. Simulations of this arrangement show that there exists a reasonably large range of stable lamellar growth, and hence we could make a comparison between simulations and the theory. We find similar agreement between our measurements and theory as we did

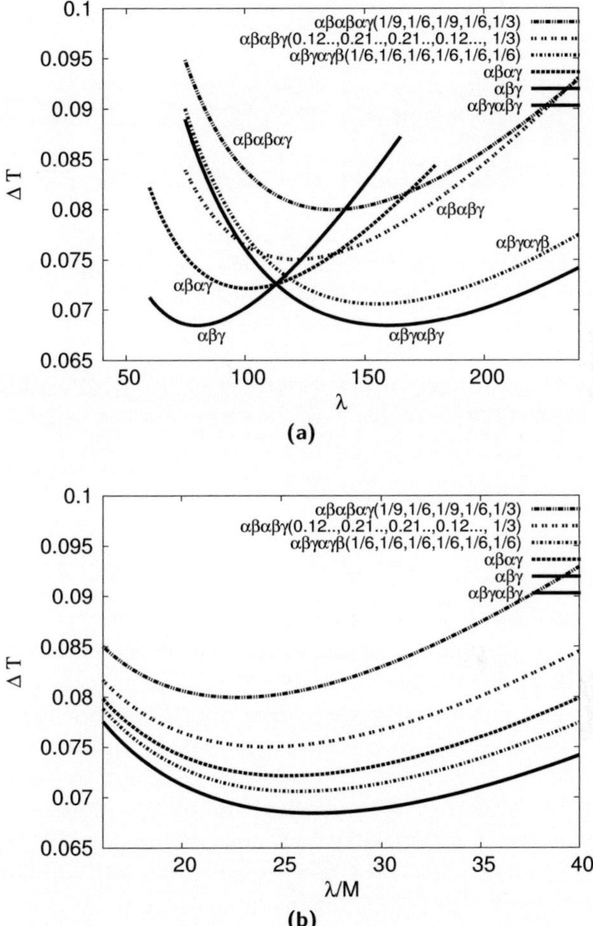


**Figure 4.21.:** (a)Synopsis of the theoretical predictions for the undercooling versus spacing of possible arrangements between period length $M = 3$ to $M = 6$, i.e. starting from $\alpha\beta\gamma$ to $\alpha\beta\gamma\alpha\beta\gamma$. (b) Same plot but with the lamellar repeat distance $\lambda$ scaled with the period length M. The variation among the arrangements is purely a result of the variation of the solutal undercooling as can be infered from the discussion in Sec. 4.2.4.

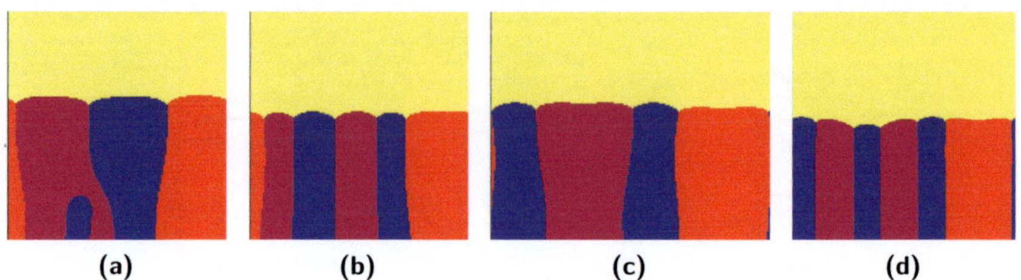

**Figure 4.22.:** Simulations starting from the arrangement $\alpha\beta\alpha\beta\alpha\gamma$ for spacings $\lambda = 135$ in a), $\lambda = 150$ in b), $\lambda = 180$ in c) and $\lambda = 240$ in d).

previously for the arrangements $\alpha\beta\gamma$ and $\alpha\beta\alpha\gamma$. The plot in Figure 4.21 shows the theoretical predictions of all the arrangements we have considered until now.

## 4.4.7. Discussion

It should by now have become clear that there exists a large number of distinct steady-state solution branches, each of which can exhibit specific instabilities. In addition, the stability thresholds potentially depend on a large number of parameters: the phase diagram data (liquidus slopes, coexistence concentration), the surface tensions (assumed identical here), and the sample concentration. Therefore, the calculation of a complete stability diagram that would generalize the one for binary eutectics of Ref. [51] represents a formidable task that is outside the scope of the present paper. Nevertheless, we can deduce from our simulations a few guidelines that can be useful for future investigation.

Lamellar steady-state solutions can be grouped into three classes, which respectively have (I) equal number of lamellae of all three phases (such as $\alpha\beta\gamma$ and $\alpha\beta\gamma\alpha\gamma\beta$), (II) equal number of lamellae for two phases (such as $\alpha\beta\alpha\gamma$), and (III) different numbers of lamellae for each phase.

For equal global volume fractions of each phase (as in most of our simulations), class III will have the narrowest stability ranges because of the simultaneous presence of very large and very thin lamella in the same arrangement, which make these patterns prone to both oscillatory and lamella elimination instabilities.

Any cycle in which a phase appears more than once can transit to another, simpler one by eliminating one lamella of this phase. This lamella instability always appears for low spacings below a critical value of the spacing that depends on the cycle. The possibility of a transition, however, depends also on the symmetries of the pattern. For instance, the arrangement $\alpha\beta\alpha\beta\alpha\gamma$, if unstable, can transform into the $\alpha\beta\gamma$, $\alpha\beta\alpha\gamma$ or the $\alpha\beta\alpha\beta\gamma$ arrangements, while for an arrangement $\alpha\beta\alpha\beta\gamma$, it is impossible to evolve into the $\alpha\beta\alpha\gamma$ arrangement if the symmetry of the pattern is preserved by the dynamics.

For large spacings, oscillatory instabilities occur and can lead to the emergence of saturated oscillatory patterns of various structures. The symmetries of the steady states seem to determine the structure of these oscillations, but no thorough survey of all possible nonlinear states was carried out.

## 4.5. Some remarks on pattern selection

Up to now, we have investigated various regular periodic patterns and their instabilities. The question which, if any, of these different arrangements, is favored for given growth conditions, is still open. From the results presented above, we can already conclude that this question cannot be answered solely on the basis of the undercooling-vs-spacing curves. Indeed, we have shown that by appropriately choosing the initial conditions, any stable configuration can be reached, regardless of its undercooling. This is also consistent with experiments and simulations on binary eutectics [34, 86]. To get some additional insights on what happens in extended systems, we conducted some simulations of isothermal solidification where the initial condition was a random lamellar arrangement. More precisely, we initialize a large system with lamellae of width $\lambda = 25$ and choose a random sequence of phases such that two neighboring lamellae are of different phases as shown in Figure 4.23a. The global probabilities of all the phases are 1/3, which corresponds to the eutectic concentration, and the temperature is set to $T = 0.785$.

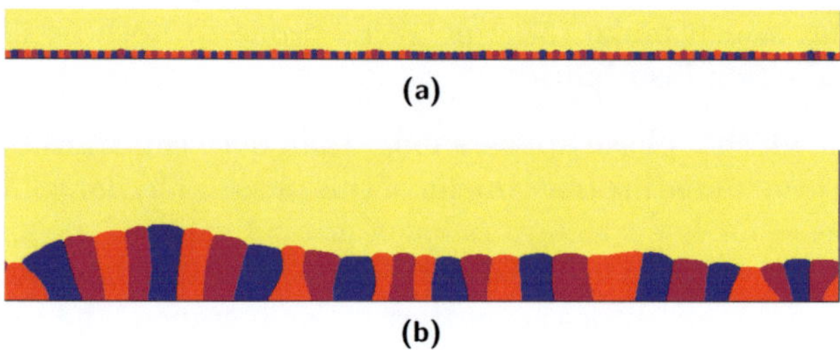

(a)

(b)

**Figure 4.23.:** Two snapshots of 2D dynamics in a large system. Isothermal simulations are started from a random configuration in (a) where the probability of occurrence of each phase is 1/3, which is also the global concentration in the liquid. The temperature of the system is T=0.785 and the concentration of the liquid is the eutectic concentration. A slowly changing pattern with a non-planar front is achieved. Some lamellae are eliminated, but no new lamellae are created.

Under isothermal growth conditions, one would expect that, at a given undercooling, the arrangement with highest local velocity would be the one that is chosen. However, in order for the front to adopt this pattern, a rearrangement of the phase sequence is necessary. In our simulations, we find that lamella elimination was possible (and indeed readily occurred). In contrast, there is no mechanism for the creation of new lamellae in our model, since we did not include fluctuations that could lead to nucleation, and the model has no spinodal decomposition that could lead to the spontaneous formation of new lamellae, as in Ref. [93]. As a result, some of the lamellae became very large in our simulations, which led to a non-planar growth front, as shown in Figure 4.23b. No clearcut periodic pattern emerged, such that our results remain inconclusive.

We believe that lamella creation is an important mechanism required for pattern adjustment. In 2D, nucleation is the only possibility for the creation of new lamellae. In contrast, in 3D, new lamellae can also form by branching mechanisms without nucleation events, since there are far more geometrical possibilities for two-phase arrangements [3, 118]. Therefore, we also conducted a few preliminary simulations in 3D.

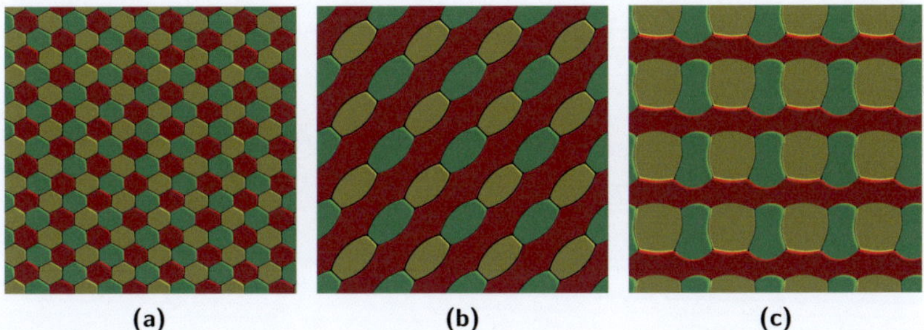

(a)         (b)         (c)

**Figure 4.24.:** Cross-sections of patterns obtained in three-dimensional directional solidification. In each picture, the simulation unit cell is tiled in a 4×4 array to get a better view of the pattern. The pattern in (a) was started from a random configuration and evolved to a perfectly hexagonal pattern (at the eutectic composition for a symmetric phase diagram). At an off-eutectic concentration, starting with two isolated rods of $\alpha$ and $\beta$ phase, the result shown in (b) is one of the possible structures, while with an asymmetric phase diagram at the eutectic concentration, we get a regular brick structure (c) from a random initial condition.

The cross sections of the simulated systems are $150 \times 150$ grid points for results in Figs. 4.24(a) and (c), and $90 \times 90$ grid points for the system Figure 4.24(b). The longest run took about 7 weeks on 80 processors, for the simulation of the pattern in Figure 4.24(a). This long simulation time is due to the fact that the pattern actually takes a long time to settle down to a steady state; the total solidification distance was of the order of 800 grid points. The other simulations required less time to reach reasonably steady states. The patterns shown in Figs. 4.24 (a) and (c) start from random initial conditions (very thin rods of randomly assigned phases), the former with the symmetric phase diagram used previously, the latter with a slightly asymmetric phase diagram constructed with the changed parameters listed below,

$$
L_i^\alpha =
\begin{pmatrix}
 & A & B & C \\
\alpha & 2.0 & 1.0 & 1.0 \\
\beta & 1.0 & 2.0 & 1.0 \\
\gamma & 1.0 & 1.0 & 2.0
\end{pmatrix}
$$

$$
T_i^\alpha =
\begin{pmatrix}
 & A & B & C \\
\alpha & 1.0 & 0.59534 & 0.63461 \\
\beta & 0.59534 & 1.0 & 0.63461 \\
\gamma & 0.59534 & 0.59534 & 1.0
\end{pmatrix}.
$$

The picture of Figure 4.24(b) corresponds to a pattern resulting of a simulation which is started with two isolated rods of $\alpha$ and $\beta$ in a matrix of $\gamma$, with an off-eutectic concentration of $c = (0.3, 0.4, 0.3)$.

As shown in Figure 4.24, many different steady-state patterns are possible in 3D. Not surprisingly, the type of pattern seen in the simulations depends on the concentration and on the phase diagram. Patterns very similar to Figure 4.24(c) have recently been observed in experiments in the Al-Ag-Cu ternary system [100]. It should be stressed that our pictures have been created by repeating the simulation cell four times in each direction in order to get a clearer view of the pattern. This means that in a larger system, the patterns might be less regular. Furthermore, we certainly have not exhausted all possible patterns. A more thorough investigation of the 3D patterns and their range of stability is left as a subject for future work.

## 4.6. Conclusion and outlook

In this chapter, we have generalized a Jackson-Hunt analysis for arbitrary periodic lamellar three-phase arrays in thin samples, and used 2D phase-field simulations to test our predictions for the minimum undercooling spacings of the various arrangements. For the model used here the value of the interface kinetic coefficient cannot be determined, which leads to some incertitude on the values of the undercooling, but this does not influence our principal findings. When the correct values of the surface free energy (that take into account additional contributions coming from the chemical part of the free energy density) are used for the comparisons with the theory, we find good agreement for the minimum undercooling spacings for all cycles investigated. Moreover, we find that, as in binary eutectics, all cycles exhibit oscillatory instabilities for spacings larger than some critical spacing. The type of oscillatory modes that are possible are determined by the set of symmetry elements of the underlying steady state.

We have repeatedly made use of symmetry arguments for a classification of the oscillatory modes. In certain cases, the symmetry is exact and general, which implies that the corresponding modes should exist for arbitrary phase diagrams and thus be observable in experiments. For instance, the mirror symmetry lines in the middle of the $\alpha$ lamellae in the $\alpha\beta\alpha\gamma$ arrangement exist even for non-symmetric phase diagrams and unequal surface tensions, and hence the corresponding oscillatory patterns and their symmetries should be universal. In other cases, we have used a symmetry element which is specific to the phase diagram used in our simulations: a mirror reflection, followed by an exchange of two phases. For a real alloy, this symmetry obviously can never be exactly realized because of asymmetries in the surface tensions, mobilities, and liquidus slopes, and therefore some of the oscillatory modes found here might not be observable in experiments. However, their occurrence cannot be completely ruled out without a detailed survey, and we expect certain characteristics to be quite robust. For instance, we have repeatedly observed that two neighboring lamellae of different phases can be interpreted as a "composite lamella" that exhibits a behavior close to the one of a single lamella in a binary eutectic pattern. Such behavior could appear even in the absence of special symmetries, and thus be generic.

Furthermore, a new type of instability (absent in binary eutectics) was found, where a cycle transforms into a simpler one by eliminating one lamella. We interpret this instability, which occurs for small spacings, through a modified version of our theoretical analysis. It is linked to the existence of an extra degree of freedom in the pattern if a given phase appears more than once in the cycle. We have not determined the full stability diagram that would be the equivalent of the one given in Ref. [51] for binary eutectics, because of the large number of independent parameters involved in the ternary problem.

We have made a few attempts to address the question of pattern selection, with inconclusive results both in 2D and 3D. In 2D, the process of pattern adjustment was hindered by the absence of a mechanism for lamella creation, and in 3D the system sizes that could be attained were too small. Based on the findings for binary eutectics, however, we believe that there is no pattern selection in the strong sense: for given processing conditions, the patterns to be found may well depend on the initial conditions and/or on the history of the system. This implies that the arrangement with the minimum undercooling may not necessarily be the one that emerges spontaneously in large-scale simulations or in experiments.

The most interesting direction of research for the future is certainly a more complete survey of pattern formation in 3D and a comparison to experimental data. To this end, either the numerical efficiency of our existing code has to be improved, or a more efficient model that generalizes the model of Ref. [32] to ternary alloys has to be developed.

# Chapter 5

# Thin interface asymptotics: Pure material solidification

# 5.1. Introduction

Phase-field modeling is an elegant technique to model a variety of problems involving phase transitions. Among them, solidification has been a field where this method has been utilized to quantitatively model a variety of microstructures involving complicated geometrical changes while evolution, both for the case of solutal and pure material problems [3, 32, 48, 50, 59, 114, 128, 134]. One of the critical points, one must take care however, is that to interpret the results of simulations in a manner which is going to be of physical value, one must have a perfect understanding as to the relation of the parameters in the equations and the free boundary problem, one is attempting to solve. This requires one to perform the asymptotics of the relevant model, which will then allow us to correctly choose the parameters in the diffuse interface description. To perform the asymptotics, two possiblities exist. One of them is the *sharp interface limit* [13], which describes, the case when the interface width tends to zero. This limit is relevant when one is performing simulations with very small interface thicknesses. However, as is often the case, in order to simulate larger microstructures, one needs to choose interface thicknesses orders of magnitude larger than the real ones. For such cases, the *thin-interface limit*, [48, 50] is more relevant as this allows us to retrieve the same free boundary, but in an easier computationally accessible manner. A fallout of this limit is that time independent free boundary problems which are relevant at low undercoolings can also be treated in a computationally efficient manner. The principal result is that when the phase evolution is coupled with that of another field, the response of the phase-field to a finite change in the coupled field can be made instantaneous in the thin-interface limit. This particular limit has been worked out for a variety of models, notably for the case of potentials which are of the smooth well type. However, the computationally more efficient double obstacle type potentials have been untreated so far. Hence, in this paper we attempt to fill this gap and make quantitative simulations possible with the use of obstacle type potentials. We derive our evolution equations from an entropy functional. We then reduce a multi-phase field model [33, 79] for treating two phase solidifications and show its equivalence to the case of single order parameter models used before, and utilize this to perform the thin-interface asymptotics. Our principal aim is to derive, the interface kinetic coefficient in the thin-interface limit and hence we simplify our analysis for the case of one dimension evolution. In addition, we treat a case where the thermal diffusivities of all the phases, and the specific heat capacities are the same.


## 5.2. Equivalence


In phase-field literature, there are two types of approaches to define order parameters. The conventional way is to use a property which varies across the interface such as density or concentration as the order parameter. The change in the property denotes the change in the microstructure or the evolution of the phase. A corollary of this approach is to use a single order parameter, which varies between two fixed limits corresponding to values in the bulk phases. While these types of models are well defined and elegant, a second approach is to define the phase-field variable as the volume fraction of the phase, which are constrained by the condition that the sum of the volume fractions of all the phases add up to 1. The evolution of the phases is efficiently tracked by the change in the volume fraction of the phases. The physical properties such as the free energy densities are then averaged among the phases using their volume fractions. As the number of phases increases, it becomes easier to track the evolution equations as the change in the volume fractions of the phases. In the following we show the equivalence of the two approaches, which will be used thereafter to perform thin-interface asymptotics for single phase pure material solidification. For the sake of discussion, we start with a free energy functional without driving forces for a two phase system, The volume fractions are denoted by $\phi_\alpha$ and $\phi_\beta$, and are constrained by the condition $\phi_\alpha + \phi_\beta = 1$. The free energy functional reads:

$$\mathcal{F}(\phi, \nabla\phi) \;=\; \int_V \left( \tilde{\sigma}_{\alpha\beta}\varepsilon \, |(\phi_\alpha \nabla\phi_\beta - \phi_\beta \nabla\phi_\alpha)|^2 + \frac{16}{\pi^2} \frac{\tilde{\sigma}_{\alpha\beta}}{\varepsilon} \phi_\alpha \phi_\beta \right) dV,$$

with surface energies $\tilde{\sigma}_{\alpha\beta}$ and $\varepsilon$ a small length scale parameter related to the width of the interface and a domain of consideration $V$. The evolution equation for the phase-field vector $\boldsymbol{\phi}$ is derived from the minimization of the free energy given by the following equation for each component,

$$\tau_{\alpha\beta}\varepsilon\frac{\partial\phi_\alpha}{\partial t} = -\frac{\delta\mathcal{F}}{\delta\phi_\alpha} - \Lambda,$$

where $\tau_{\alpha\beta}$ is the relaxation constant for the interface and $\Lambda$ is the Lagrange parameter for respecting the sum of the volume fractions of the phases as 1. Expanding the variational derivative on the right side of the equation yields,

$$\tau_{\alpha\beta}\varepsilon\frac{\partial\phi_\alpha}{\partial t} = \tilde{\sigma}_{\alpha\beta}\varepsilon\Big( -2\nabla\cdot(\phi_\beta(\phi_\alpha\nabla\phi_\beta - \phi_\beta\nabla\phi_\alpha))$$
$$- 2\nabla\phi_\beta\cdot(\phi_\alpha\nabla\phi_\beta - \phi_\beta\nabla\phi_\alpha)\Big) - \frac{16}{\pi^2}\frac{\tilde{\sigma}_{\alpha\beta}}{\varepsilon}\phi_\beta - \Lambda$$
$$= -4\tilde{\sigma}_{\alpha\beta}\varepsilon\Big(\nabla\phi_\beta\cdot(\phi_\alpha\nabla\phi_\beta - \phi_\beta\nabla\phi_\alpha) +$$
$$\frac{1}{2}\phi_\beta(\phi_\alpha\nabla^2\phi_\beta - \phi_\beta\nabla^2\phi_\alpha)\Big) - \frac{16}{\pi^2}\frac{\tilde{\sigma}_{\alpha\beta}}{\varepsilon}\phi_\beta - \Lambda.$$

Utilizing the properties $\nabla\phi_\alpha = -\nabla\phi_\beta$ and $\phi_\alpha + \phi_\beta = 1$ for the case of two phases gives,

$$\tau_{\alpha\beta}\varepsilon\frac{\partial\phi_\alpha}{\partial t} = -4\tilde{\sigma}_{\alpha\beta}\varepsilon\left(|\nabla\phi_\beta|^2 + \frac{1}{2}\phi_\beta(\nabla^2\phi_\beta)\right) - \frac{16}{\pi^2}\frac{\tilde{\sigma}_{\alpha\beta}}{\varepsilon}\phi_\beta - \Lambda,$$

and the Lagrange parameter $\Lambda$ as,

$$\Lambda = -2\tilde{\sigma}_{\alpha\beta}\varepsilon\left(|\nabla\phi_\beta|^2 + |\nabla\phi_\alpha|^2\right) - \tilde{\sigma}_{\alpha\beta}\varepsilon\left(\phi_\beta\nabla^2\phi_\beta + \phi_\alpha\nabla^2\phi_\alpha\right) -$$
$$\frac{16}{\pi^2}\frac{\tilde{\sigma}_{\alpha\beta}}{2\varepsilon}(\phi_\beta + \phi_\alpha).$$

Including the Lagrange parameter, the evolution equation for a two phase system transforms to,

$$\tau_{\alpha\beta}\varepsilon\frac{\partial\phi_\alpha}{\partial t} = \tilde{\sigma}_{\alpha\beta}\varepsilon\nabla^2\phi_\alpha - \frac{16}{\pi^2}\frac{\tilde{\sigma}_{\alpha\beta}}{2\varepsilon}(1 - 2\phi_\alpha). \tag{5.1}$$

Equivalently, the evolution equation for the $\beta$ phase reads,

$$\tau_{\alpha\beta}\varepsilon\frac{\partial\phi_\beta}{\partial t} \;=\; \tilde{\sigma}_{\alpha\beta}\varepsilon\nabla^2\phi_\beta - \frac{16}{\pi^2}\frac{\tilde{\sigma}_{\alpha\beta}}{2\varepsilon}\left(1 - 2\phi_\beta\right). \tag{5.2}$$

With the construction of the Lagrange parameter, the sum of the evolution equations for the two phases is zero. Hence, there exists only *one independent equation* in the system and one can choose either to derive the dynamics of the system of two phases. With no loss of generality, we choose the evolution equation of the $\alpha$ phase for further discussion and analysis. The equilibrium condition between two phases is given by the condition $\dfrac{\partial\phi_\alpha}{\partial t} = 0$, which implies,

$$\tilde{\sigma}_{\alpha\beta}\varepsilon\nabla^2\phi_\alpha \;=\; \frac{16}{\pi^2}\frac{\tilde{\sigma}_{\alpha\beta}}{2\varepsilon}\left(1 - 2\phi_\alpha\right).$$

As we are treating two-phase interfaces, the surface energy $\tilde{\sigma}_{\alpha\beta}$ is binary interface property completely defined by the sharp interface free boundary problem. However, in the special case when there are adsorbed phases at the interface, one must compute the surface energy numerically from the equilibrated profiles of the phases. The treatment however, remains in principle similar to the case of binary interfaces, where we find the extremum of the free energy per unit area. The surface energy is defined as integral of the free-energy functional per-unit area computed using the phase-field function $\phi_\alpha\left(x\right)$ which maximizes the free energy. Equivalently, this is the solution to the equilibrium phase-field equation in 1D. The resulting solution of the partial diffusion equation are the equilibrium phase field profiles which can be substituted back into the functional and integrated to get the surface energy. In 1D, the equilibrium equation is multiplied by $\dfrac{\partial\phi_\alpha}{\partial x}$, on both sides and integrated from 0 to $x$. We get,

$$\tilde{\sigma}_{\alpha\beta}\varepsilon\left(\frac{d\phi_\alpha}{dx}\right)^2 = \frac{16}{\pi^2}\frac{\tilde{\sigma}_{\alpha\beta}}{\varepsilon}\phi_\alpha\left(1 - \phi_\alpha\right). \tag{5.3}$$

Equivalently, a similar expression can be formulated for the $\beta$ phase,

$$\tilde{\sigma}_{\alpha\beta}\varepsilon\left(\frac{d\phi_\beta}{dx}\right)^2 = \frac{16}{\pi^2}\frac{\tilde{\sigma}_{\alpha\beta}}{\varepsilon}\phi_\beta\left(1 - \phi_\beta\right).$$

The slope of the individual profiles can be derived as,

$$\frac{d\phi_\alpha}{dx} = \pm\frac{1}{\varepsilon}\sqrt{\frac{16}{\pi^2}\phi_\alpha\left(1-\phi_\alpha\right)},$$

$$\frac{d\phi_\beta}{dx} = \pm\frac{1}{\varepsilon}\sqrt{\frac{16}{\pi^2}\phi_\beta\left(1-\phi_\beta\right)}.$$

The sign of the derivative of each profile of $\alpha$ and the $\beta$ phases is derived from the boundary conditions. It depends on the direction from $\phi_\alpha = 0$ to $\phi_\alpha = 1$ as $x$ goes from $-\infty$ to $+\infty$ or the other way around. In the following we choose the signs of the derivatives as,

$$\frac{d\phi_\alpha}{dx} = +\frac{1}{\varepsilon}\sqrt{\frac{16}{\pi^2}\phi_\alpha\left(1-\phi_\alpha\right)}, \tag{5.4}$$

$$\frac{d\phi_\beta}{dx} = -\frac{1}{\varepsilon}\sqrt{\frac{16}{\pi^2}\phi_\beta\left(1-\phi_\beta\right)}. \tag{5.5}$$

Next, we recollect the gradient potential of the form, $\tilde{\sigma}_{\alpha\beta}\varepsilon|\phi_\alpha\nabla\phi_\beta - \phi_\beta\nabla\phi_\alpha|^2$, and substitute the form of the gradients $\dfrac{d\phi_\alpha}{dx}$, Eqn. (5.4) and $\dfrac{d\phi_\beta}{dx}$, Eqn. (5.5) and using $\dfrac{d\phi_\beta}{dx} = -\dfrac{d\phi_\beta}{dx}$, we get $\tilde{\sigma}_{\alpha\beta}\varepsilon|\phi_\alpha\nabla\phi_\beta - \phi_\beta\nabla\phi_\alpha|^2 = \dfrac{16}{\pi^2}\dfrac{\tilde{\sigma}_{\alpha\beta}\phi_\alpha\left(1-\phi_\alpha\right)}{\varepsilon}$ and hence the surface energy is given by,

$$\overline{\sigma_{\alpha\beta}} = 2\tilde{\sigma}_{\alpha\beta}\int_X \frac{16}{\pi^2}\frac{\phi_\alpha\left(1-\phi_\alpha\right)}{\varepsilon}dx.$$

By changing the variables from $x$ to $\phi_\alpha$ and applying the relation in Eqn. (5.3), the equivalent expression for the surface energy derives,

$$\overline{\sigma_{\alpha\beta}} = 2\tilde{\sigma}_{\alpha\beta}\int_0^1 \sqrt{\left(\frac{16}{\pi^2}\phi_\alpha\left(1-\phi_\alpha\right)\right)}d\phi_\alpha.$$

The above integral has been constructed in such a way so as to return $1/2$ i.e. the reason for the choice of the factor $16/(\pi^2)$, such that the surface energy $\overline{\sigma_{\alpha\beta}}$ is $\tilde{\sigma}_{\alpha\beta}$. The interface thickness can similarly be calculated by using Eqn. (5.3) and integrated such that,

$$\tilde{\Lambda}_{\alpha\beta} \;=\; \int_0^1 \frac{\varepsilon d\phi_\alpha}{\sqrt{\dfrac{16}{\pi^2}\phi_\alpha\left(1-\phi_\alpha\right))}},$$

which computes as $\tilde{\Lambda}_{\alpha\beta} = \dfrac{\pi^2\varepsilon}{4} \approx 2.5\varepsilon$.

One can also use the difference of the two equations i.e Eqn. $(5.1) - (5.2)$ as the independent equation, describing the evolution of the system as,

$$\tau_{\alpha\beta}\varepsilon\frac{\partial\left(\phi_\alpha-\phi_\beta\right)}{\partial t} \;=\; \tilde{\sigma}_{\alpha\beta}\varepsilon\nabla^2\left(\phi_\alpha-\phi_\beta\right) - \frac{16}{\pi^2}\frac{\tilde{\sigma}_{\alpha\beta}}{2\varepsilon}\left(1-2\phi_\alpha-\left(1-2\phi_\beta\right)\right).$$

Using $\phi_\alpha + \phi_\beta = 1$, and simplifying we arrive at,

$$2\tau_{\alpha\beta}\varepsilon\frac{\partial\phi_\alpha}{\partial t} \;=\; 2\tilde{\sigma}_{\alpha\beta}\varepsilon\nabla^2\phi_\alpha - \frac{16}{\pi^2}\frac{\tilde{\sigma}_{\alpha\beta}}{\varepsilon}\left(1-2\phi_\alpha\right).$$

This is the same equation we obtain, by starting from a one order parameter description of the model. Dividing throughout by 2, we get the identical expression as the evolution equation for the phases, described in Eqn.(5.1) or Eqn.(5.2). This establishes the equivalence of the method to one-order parameter models.

## 5.3. Asymptotic analysis

In the following, we derive the thin interface corrections for the case of a pure material solidification. We perform the expansions of the phase-field variable $\phi_\alpha$ and temperature field $T$ as powers of 'p' which is a small parameter called the *interface Peclet number* $Pe = \dfrac{W}{(D/V)}$, where $W$ is the interface width and $D/V$ is the diffusion length in the problem. To derive the evolution equations, we start from an entropy functional which is elaborated as,

$$\mathcal{S} = \int_V s\left(e, \phi\right) - \left(\varepsilon a\left(\phi, \nabla\phi\right) + \frac{1}{\varepsilon} w\left(\phi\right)\right) dV,$$

where, $e$ is the internal energy of the system, $a$ is the gradient entropy density and $w$ is the surface potential density. The evolution equation for the internal energy can be derived as a conservation law as follows,

$$\frac{\partial e}{\partial t} = -\nabla \cdot J_e \quad \text{with} \quad J_e = -M\left(\phi\right)\nabla\frac{\delta\mathcal{S}}{\delta e},$$

where $J_e$ is the flux of the internal energy, and $M$ is mobility related to this flux. The variational derivative $\dfrac{\delta\mathcal{S}}{\delta e}$ can be derived as $\dfrac{1}{T}$. Substituting this result in the preceding equation, the evolution equation for the internal energy transforms to,

$$\frac{\partial e}{\partial t} = \nabla \cdot \left(M\left(\phi\right)\nabla\frac{1}{T}\right).$$

Applying the relation $e = f + TS$, where $f$ is the Helmholtz free energy density of the system, we derive the evolution equation for the temperature field as,

$$\frac{\partial T}{\partial t} = \frac{\nabla \cdot \left(M\left(\phi\right)\nabla\frac{1}{T}\right) - \sum_{\alpha=1}^{N} \frac{\partial e}{\partial \phi_\alpha}\frac{\partial \phi_\alpha}{\partial t}}{-T\frac{\partial^2 f}{\partial T^2}}.$$

For treating problems close to the melting point $T_m$, we can derive the free energy of each phase as $\left(\dfrac{L_\alpha\left(T - T_m^\alpha\right)}{T_m^\alpha} + C_v^\alpha\left(T - T_m^\alpha\right)\right)$. It follows, that the internal energy of each phase can be written as $e^\alpha = L_\alpha + C_v^\alpha\left(T - T_m^\alpha\right)$, where $C_v^\alpha$ is the volumetric heat capacity at constant volume of the $\alpha$ phase and $L_\alpha$ is the latent heat of fusion of the $\alpha$ phase. The total internal energy of the system can be elaborated as $e\left(\phi, T\right) = \sum_{\alpha=1}^{N} e^\alpha h_\alpha\left(\phi\right)$, where $h_\alpha\left(\phi\right)$ is the interpolation function. We make another simplification, that, for the case we have all the $C_v^\alpha$ as equal, we can derive the following evolution equation for the temperature field,

$$\frac{\partial T}{\partial t} = \frac{\nabla \cdot \left( M\left(\phi\right) \nabla \frac{1}{T} \right) + \sum_{\alpha=1}^{N} L_\alpha \frac{\partial h_\alpha\left(\phi\right)}{\partial \phi_\alpha} \frac{\partial \phi_\alpha}{\partial t}}{-T \frac{\partial^2 f}{\partial T^2}},$$

which, upon employing the thermodynamic relation $-T \frac{\partial^2 f}{\partial T^2} = C_v$ becomes,

$$\frac{\partial T}{\partial t} = \frac{\nabla \cdot \left( M\left(\phi\right) \nabla \frac{1}{T} \right) + \sum_{\alpha=1}^{N} L_\alpha \frac{\partial h_\alpha\left(\phi\right)}{\partial \phi_\alpha} \frac{\partial \phi_\alpha}{\partial t}}{C_v}.$$

In the above, we write $M\left(\phi\right) = \sum_{\alpha=1}^{N} M^\alpha h_\alpha\left(\phi_\alpha\right)$ and assume the free energies with respect to the liquid as the reference which implies that the latent heats of fusion $L_\alpha$ are non-zero only for the solid phases. The evolution equation of the phase-field variables $\phi = \{\phi_\alpha\}_{\alpha=1}^{N}$ are derived from the phenomenological maximization of the entropy functional,

$$\omega_{\alpha\beta}\varepsilon \frac{\partial \phi_\alpha}{\partial t} = \frac{\delta S}{\delta \phi_\alpha} - \Lambda,$$

where $\omega_{\alpha\beta}$ is the constant related to the relaxation of the phase-field, derived from the entropy functional. We now reduce our system to two phases, a pure solid $\alpha$ and a pure liquid $\beta$. We employ interpolation functions of the form $h_\alpha\left(\phi\right) = h_\alpha\left(\phi_\alpha\right)$. Defining the mobility, $M^\alpha = -K^\alpha T^2$ for each phase $\alpha$ and utilizing the discussion in section 5.2, we write a single independent evolution equation of the order parameter $\phi_\alpha$ as follows,

$$\omega_{\alpha\beta}\varepsilon \frac{\partial \phi_\alpha}{\partial t} = -4\gamma_{\alpha\beta}\varepsilon \left( |\nabla \phi_\beta|^2 + \frac{1}{2}\phi_\beta \left(\nabla^2 \phi_\beta\right) \right) - \frac{16}{\pi^2} \frac{\gamma_{\alpha\beta}}{\varepsilon} \phi_\beta$$
$$- \frac{f_\alpha\left(T\right)}{T} \frac{\partial h_\alpha\left(\phi_\alpha\right)}{\partial \phi_\alpha} - \Lambda,$$
$$C_v \frac{\partial T}{\partial t} = \nabla \cdot \left( K\left(\phi\right) \nabla T \right) + L_\alpha \frac{\partial h_\alpha\left(\phi_\alpha\right)}{\partial t},$$

where we have written the thermal conductivity as $K\left(\boldsymbol{\phi}\right) = \sum_{\alpha=1}^{N} K^{\alpha} h_{\alpha}\left(\phi_{\alpha}\right)$.

Further, we used $\dfrac{\delta s}{\delta \phi_{\alpha}} = -\dfrac{f_{\alpha}\left(T\right)}{T} \dfrac{\partial h_{\alpha}\left(\phi_{\alpha}\right)}{\partial \phi_{\alpha}}$. The parameter $\gamma_{\alpha\beta}$ denotes the surface entropy density. For the discussion hereafter, the thermal conductivities are assumed the same for both phases, given by $K$. The Lagrange parameter $\Lambda$ is expanded as,

$$\Lambda = -2\gamma_{\alpha\beta}\varepsilon \left(\left|\nabla\phi_{\beta}\right|^{2} + \left|\nabla\phi_{\alpha}\right|^{2}\right) - \gamma_{\alpha\beta}\varepsilon \left(\phi_{\beta}\nabla^{2}\phi_{\beta} + \phi_{\alpha}\nabla^{2}\phi_{\alpha}\right) -$$
$$\frac{16}{\pi^{2}} \frac{\gamma_{\alpha\beta}}{2\varepsilon} \left(\phi_{\beta} + \phi_{\alpha}\right) - \frac{1}{2T}\left(f_{\alpha}\left(T\right)\frac{\partial h_{\alpha}\left(\phi_{\alpha}\right)}{\partial \phi_{\alpha}} + f_{\beta}\left(T\right)\frac{\partial h_{\beta}\left(\phi_{\beta}\right)}{\partial \phi_{\beta}}\right),$$

which upon incorporation in the evolution equation and with the assumption $f_{\beta}\left(T\right) = 0$ (because $L_{\beta} = 0$) reads as,

$$\omega_{\alpha\beta}\varepsilon\frac{\partial\phi_{\alpha}}{\partial t} = \gamma_{\alpha\beta}\varepsilon\nabla^{2}\phi_{\alpha} - \frac{16}{\pi^{2}}\frac{\gamma_{\alpha\beta}}{2\varepsilon}\left(1 - 2\phi_{\alpha}\right) - \frac{1}{2}\frac{L_{\alpha}\left(T - T_{m}^{\alpha}\right)}{T T_{m}^{\alpha}}\frac{\partial h_{\alpha}\left(\phi_{\alpha}\right)}{\partial\phi_{\alpha}},$$
$$\tag{5.6}$$

$$C_{v}\frac{\partial T}{\partial t} = \nabla \cdot \left(K\nabla T\right) + L_{\alpha}\frac{\partial h_{\alpha}\left(\phi_{\alpha}\right)}{\partial t}.$$

Multiplying Eqn. (5.6) by $T$, we get,

$$\tau_{\alpha\beta}\varepsilon\frac{\partial\phi_{\alpha}}{\partial t} = \tilde{\sigma}_{\alpha\beta}\varepsilon\nabla^{2}\phi_{\alpha} - \frac{16}{\pi^{2}}\frac{\tilde{\sigma}_{\alpha\beta}}{2\varepsilon}\left(1 - 2\phi_{\alpha}\right) - \frac{1}{2}\frac{L_{\alpha}\left(T - T_{m}^{\alpha}\right)}{T_{m}^{\alpha}}\frac{\partial h_{\alpha}\left(\phi_{\alpha}\right)}{\partial\phi_{\alpha}},$$

where $\tau_{\alpha\beta} = T\omega_{\alpha\beta}$ and $\tilde{\sigma}_{\alpha\beta} = T\gamma_{\alpha\beta}$. We will use this evolution equation henceforth in the analysis. We next non-dimensionalize the equations using the diffusion length $l_{c}$ as the length scale and use the time scale $l_{c}^{2}/\kappa$. Here $\kappa$ is the thermal diffusivity which is defined as $K/C_{v}$. We introduce a small parameter $p = \varepsilon/l_{c}$ and the re-scaled equations are rewritten in 1D as,

$$\tilde{\tau}p^{2}\frac{\partial\phi_{\alpha}}{\partial t} \quad = \quad p^{2}\frac{\partial^{2}\phi_{\alpha}}{\partial x^{2}} - \frac{16}{2\pi^{2}}\left(1 - 2\phi_{\alpha}\right) - \tilde{\alpha}\frac{p}{2}\frac{T - T_{m}^{\alpha}}{T_{m}^{\alpha}}\frac{\partial h_{\alpha}\left(\phi_{\alpha}\right)}{\partial\phi_{\alpha}} \tag{5.7}$$

$$\frac{\partial T}{\partial t} \quad = \quad \frac{\partial^{2}T}{\partial x^{2}} + \tilde{\lambda}\frac{\partial h_{\alpha}\left(\phi_{\alpha}\right)}{\partial t}, \tag{5.8}$$

where we have additionally defined the parameters $\tilde{\lambda} = \dfrac{L_\alpha}{C_v}$, $\tilde{\alpha} = \dfrac{l_c}{(\tilde{\sigma}_{\alpha\beta}/L_\alpha)}$ and $\tilde{\tau} = \tau_{\alpha\beta}\kappa/\tilde{\sigma}_{\alpha\beta}$. In Eqn. (5.8), the physical situation remains unchanged if we calculate the temperature changes with respect to the melting temperature $T_m^\alpha$, such that we can transform the variable to $(T - T_m^\alpha)$. Thereafter, we divide the whole equation by $\tilde{\lambda}$, giving,

$$\frac{\partial u}{\partial t} = \frac{\partial^2 u}{\partial x^2} + \frac{\partial h_\alpha\left(\phi_\alpha\right)}{\partial t},$$

where $u = \dfrac{T - T_m^\alpha}{(L_\alpha/C_v)}$. With this substitution, and setting $g = \tilde{\alpha}L_\alpha/(C_v T_m^\alpha)$, we get the following evolution equation for $\phi_\alpha$,

$$\tilde{\tau}p^2\frac{\partial \phi_\alpha}{\partial t} = p^2\frac{\partial^2 \phi_\alpha}{\partial x^2} - \frac{16}{2\pi^2}\left(1 - 2\phi_\alpha\right) - \frac{gup}{2}\frac{\partial h_\alpha\left(\phi_\alpha\right)}{\partial \phi_\alpha}.$$

Next we transform the equations in the moving frame at steady-state, such that the total derivative with respect to time vanishes. We get,

$$-v\tilde{\tau}p^2\frac{\partial \phi_\alpha}{\partial x} = p^2\frac{\partial^2 \phi_\alpha}{\partial x^2} - \frac{16}{2\pi^2}\left(1 - 2\phi_\alpha\right) - \frac{gup}{2}\frac{\partial h_\alpha\left(\phi_\alpha\right)}{\partial \phi_\alpha}, \quad (5.9)$$

$$-v\frac{\partial u}{\partial x} = \frac{\partial^2 u}{\partial x^2} - v\frac{\partial h_\alpha\left(\phi_\alpha\right)}{\partial t}, \quad (5.10)$$

where the velocity $(V)$ is measured in the non-dimensional transformed co-ordinate set as $v = Vl_c/\kappa$. There are two regions in the spatial solutions of $\phi$ and $u$: The outer region(bulk solid and liquid) where there is slow change in the variables and an inner region where there is rapid change. To probe into the inner solutions, we scale our co-ordinate with respect to the interface Peclet number $p$, by introducing a scaled co-ordinate system $\eta = x/p$. With this transformation, the evolution equations become,

$$-\tilde{\tau}pv\frac{\partial \phi_\alpha}{\partial \eta} = \frac{\partial^2 \phi_\alpha}{\partial \eta^2} - \frac{16}{2\pi^2}\left(1 - 2\phi_\alpha\right) - \frac{gup}{2}\frac{\partial h_\alpha\left(\phi_\alpha\right)}{\partial \phi_\alpha},$$

$$-\frac{v}{p}\frac{\partial u}{\partial \eta} = \frac{1}{p^2}\frac{\partial^2 u}{\partial x^2} - \frac{v}{p}\frac{\partial h_\alpha\left(\phi_\alpha\right)}{\partial t}.$$

The outer and inner solutions are written as expansions of the parameter $p$. The outer solutions are denoted as $\widetilde{\phi}_\alpha = \widetilde{\phi^0_\alpha} + p\widetilde{\phi^1_\alpha} + p^2\widetilde{\phi^2_\alpha}\ldots$ and $\widetilde{u} = \widetilde{u^0} + p\widetilde{u^1} + p^2\widetilde{u^2}\ldots$, while the inner solutions are $\phi_\alpha = \phi^0_\alpha + p\phi^1_\alpha + p^2\phi^2_\alpha\ldots$ and $u = u^0 + pu^1 + p^2u^2\ldots$. The outer solution is $\widetilde{\phi}_\alpha = 0,1$ in the bulk liquid or solid respectively and is stable to any order in $p$. The outer solution for the temperature satisfies the diffusion equation,

$$\frac{\partial \widetilde{u}}{\partial t} = \frac{\partial^2 \widetilde{u}}{\partial x^2}.$$

Given this, the matching conditions for $\phi_\alpha$ are trivial while for the $u$ field the matching conditions are derived by comparing the equations order by order in $p$ as,

$$\lim_{\eta\to\pm\infty} u^0 = \widetilde{u^0}\Big|^{\pm} \tag{5.11}$$

$$\lim_{\eta\to\pm\infty} u^1 = \lim_{\eta\to\pm\infty}\left(\widetilde{u^1}\Big|^{\pm} + \eta\frac{\partial\widetilde{u^0}}{\partial x}\Big|^{\pm}\right) \tag{5.12}$$

$$\lim_{\eta\to\pm\infty} u^2 = \lim_{\eta\to\pm\infty}\left(\widetilde{u^2}\Big|^{\pm} + \eta\frac{\partial\widetilde{u^1}}{\partial x}\Big|^{\pm} + \frac{\eta^2}{2}\frac{\partial^2\widetilde{u^0}}{\partial x^2}\Big|^{\pm}\right) \tag{5.13}$$

and, the derivative matching conditions,

$$\lim_{\eta\to\pm\infty}\frac{\partial u^0}{\partial\eta} = 0 \tag{5.14}$$

$$\lim_{\eta\to\pm\infty}\frac{\partial u^1}{\partial\eta} = \frac{\partial\widetilde{u^0}}{\partial x}\Big|^{\pm} \tag{5.15}$$

$$\lim_{\eta\to\pm\infty}\frac{\partial u^2}{\partial\eta} = \lim_{\eta\to\pm\infty}\left(\frac{\partial\widetilde{u^1}}{\partial x}\Big|^{\pm} + \eta\frac{\partial^2\widetilde{u^0}}{\partial x^2}\Big|^{\pm}\right). \tag{5.16}$$

Next we proceed to write the evolution equation of $\phi_\alpha$ and $u$ order by order in p and derive the relevant boundary conditions as solvability conditions. The phase-field equation at order $p^0$ writes,

$$\frac{\partial^2\phi^0_\alpha}{\partial\eta^2} = \frac{16}{2\pi^2}\left(1 - 2\phi^0_\alpha\right).$$

Multiplying with $\dfrac{\partial \phi_\alpha^0}{\partial x}$ on both sides and integrating we get,

$$\frac{1}{2}\left(\frac{\partial \phi_\alpha^0}{\partial \eta}\right)^2 = \frac{1}{2}\frac{16}{\pi^2}\phi_\alpha^0\left(1-\phi_\alpha^0\right)$$

$$\frac{\partial \phi_\alpha^0}{\partial \eta} = \pm\frac{4}{\pi}\sqrt{\phi_\alpha^0\left(1-\phi_\alpha^0\right)}.$$

With the boundary conditions $\phi_\alpha = 1(\text{solid})$ at $x = -\infty$ and $\phi_\alpha = 0(\text{liquid})$ at $x = +\infty$ we have,

$$\frac{\partial \phi_\alpha^0}{\partial \eta} = -\frac{4}{\pi}\sqrt{\phi_\alpha^0\left(1-\phi_\alpha^0\right)}. \tag{5.17}$$

The $u$ equation at order $1/p^2$ is,

$$\frac{\partial^2 u_0}{\partial \eta^2} = 0.$$

which upon integrating once becomes,

$$\frac{\partial u_0}{\partial \eta} = A_1.$$

From the matching condition in Eqn. (5.14) we obtain $\lim_{\eta\to\pm\infty} \partial_\eta u^0 = 0$, which implies $A_1 = 0$, which upon subsequent integration reads,

$$u_0 = \overline{u_0},$$

with integration constant $\overline{u_0}$. In order to determine this constant, we substitute $u_0$ into the phase-field equation at order $p^1$ which gives,

$$-\tilde{\tau}pv\frac{\partial \phi_\alpha^0}{\partial \eta} = p\frac{\partial^2 \phi_\alpha^1}{\partial \eta^2} + \frac{16}{\pi^2}p\phi_\alpha^1 - p\frac{gu^0}{2}\frac{\partial h\left(\phi_\alpha^0\right)}{\partial \phi_\alpha},$$

and thereafter simplifies to,

$$-\tilde{\tau}v\frac{\partial \phi_\alpha^0}{\partial \eta} = \frac{\partial^2 \phi_\alpha^1}{\partial \eta^2} + \frac{16}{\pi^2}\phi_\alpha^1 - \frac{gu^0}{2}\frac{\partial h\left(\phi_\alpha^0\right)}{\partial \phi_\alpha}.$$

*Note, the term* $\dfrac{16}{\pi^2}$ *corresponds to* $-\dfrac{16}{2\pi^2}\dfrac{\partial}{\partial\phi_\alpha}\left(1-2\phi_\alpha\right)$ *which is one of the terms in the expansion of the potential function in increasing orders of the interface Peclet number* $p$. *This can be written for any function of* $\phi_\alpha$, *e.g* $s\left(\phi_\alpha\right)$ *by substituting expansion of* $\phi_\alpha$ *in orders of* $p$ *as.*

$$s\left(\phi_\alpha^0 + p\phi_\alpha^1 + p^2\phi_\alpha^2\right) = s\left(\phi_\alpha^0\right) + p\frac{\partial s\left(\phi_\alpha^0\right)}{\partial\phi_\alpha}\phi_\alpha^1 + p^2\frac{\partial s\left(\phi_\alpha^0\right)}{\partial\phi_\alpha}\phi_\alpha^2 +$$

$$\frac{p^2}{2}\frac{\partial^2 s\left(\phi_\alpha^0\right)}{\partial\phi_\alpha^2}\left(\phi_\alpha^1\right)^2 \ldots,$$

*where we limit the above expansion to order* $p^2$. Here it is useful to identify an operator which derives from the phase-field equation at order $p^0$ as follows,

$$\frac{\partial^2\phi_\alpha^0}{\partial\eta^2} - \frac{16}{2\pi^2}\left(1 - 2\phi_\alpha^0\right) = 0.$$

Differentiating the above once, gives:

$$\frac{\partial^3\phi_\alpha^0}{\partial\eta^3} + \frac{16}{\pi^2}\frac{\partial\phi_\alpha^0}{\partial\eta} = 0.$$

Written as,

$$\left(\frac{\partial^2}{\partial\eta^2} + \frac{16}{\pi^2}\right)\frac{\partial\phi_\alpha^0}{\partial\eta} = 0,$$

implies that the term in the brackets $\dfrac{\partial^2}{\partial\eta^2} + \dfrac{16}{\pi^2}$ can be defined as a linear operator denoted by $L$. We also derive that $\partial_\eta\phi_\alpha^0$ is a homogeneous solution of the operator $L$. Writing the phase-field equation at order $p^1$ by employing the operator $L$ gives,

$$L\phi_\alpha^1 = -\tilde{\tau}v\frac{\partial\phi_\alpha^0}{\partial\eta} + \frac{gu^0}{2}\frac{\partial h\left(\phi_\alpha^0\right)}{\partial\phi_\alpha}.$$

Multiplying both sides with $\dfrac{\partial \phi_\alpha^0}{\partial \eta}$ and integrating we get,

$$\int_{-\infty}^{\infty} L\phi_\alpha^1 \frac{\partial \phi_\alpha^0}{\partial \eta} \partial \eta \;=\; -\tilde{\tau}v \int_{-\infty}^{\infty} \left(\frac{\partial \phi_\alpha^0}{\partial \eta}\right)^2 \partial \eta + \int_{-\infty}^{\infty} \frac{gu^0}{2} \frac{\partial h\left(\phi_\alpha^0\right)}{\partial \phi_\alpha} \frac{\partial \phi_\alpha^0}{\partial \eta} \partial \eta$$

The left hand side of this equation can be expanded as,

$$\int_{-\infty}^{\infty} L\phi_\alpha^1 \frac{\partial \phi_\alpha^0}{\partial \eta} \partial \eta = \int_{-\infty}^{\infty} \frac{\partial^2 \phi_\alpha^1}{\partial \eta^2} \frac{\partial \phi_\alpha^0}{\partial \eta} \partial \eta + \int_{-\infty}^{\infty} \frac{16}{\pi^2} \phi_\alpha^1 \frac{\partial \phi_\alpha^0}{\partial \eta} \partial \eta, \qquad (5.18)$$

which upon partial integration becomes,

$$= \left. \frac{\partial \phi_\alpha^1}{\partial \eta} \frac{\partial \phi_\alpha^0}{\partial \eta} \right|_{-\infty}^{\infty} - \int_{-\infty}^{\infty} \frac{\partial \phi_\alpha^1}{\partial \eta} \frac{\partial^2 \phi_\alpha^0}{\partial \eta^2} \partial \eta$$

$$+ \left. \frac{16}{2\pi^2} \phi_\alpha^1 \left(2\phi_\alpha^0 - 1\right) \right|_{-\infty}^{\infty} + \int_{-\infty}^{\infty} \frac{16}{2\pi^2} \frac{\partial \phi_\alpha^1}{\partial \eta} \left(1 - 2\phi_\alpha^0\right) \partial \eta.$$

*While performing the integrations and substitutions of functions it is necessary to take care that the reference point is at $\eta = 0$, where $\phi_\alpha = 0.5$. This is because the moving frame is fixed to this point.* The two constants are zero owing to $\phi_\alpha^1 \to 0$ at both extremities along with the derivatives which implies that the integral simplifies to,

$$\int_{-\infty}^{\infty} L\phi_\alpha^1 \frac{\partial \phi_\alpha^0}{\partial \eta} \partial \eta \;=\; -\int_{-\infty}^{\infty} \left(\frac{\partial^2 \phi_\alpha^0}{\partial \eta^2} - \frac{16}{2\pi^2}\left(1 - 2\phi_\alpha^0\right)\right) \frac{\partial \phi_\alpha^1}{\partial \eta} \partial \eta.$$

The term inside the brackets is the phase-field equation at order $p^0$ and hence is zero, giving the first solvability condition,

$$-\tilde{\tau}v \int_{-\infty}^{\infty} \left(\frac{\partial \phi_\alpha^0}{\partial \eta}\right)^2 \partial \eta + \int_{-\infty}^{\infty} \frac{gu^0}{2} \frac{\partial h\left(\phi_\alpha^0\right)}{\partial \phi_\alpha} \frac{\partial \phi_\alpha^0}{\partial \eta} \partial \eta \;=\; 0. \quad (5.19)$$

For the obstacle potential and any interpolation polynomial $h_\alpha\left(\phi_\alpha\right)$ varying from 0 to 1, the two integrals in the above solvability condition are easily computed as,

$$\int_{-\infty}^{\infty} \left(\frac{\partial \phi_\alpha^0}{\partial \eta}\right)^2 \partial \eta = \int_{1}^{0} \frac{\partial \phi_\alpha^0}{\partial \eta} \partial \phi_\alpha = \frac{4}{\pi} \int_{0}^{1} \sqrt{\phi_\alpha^0 \left(1 - \phi_\alpha^0\right)} \partial \phi_\alpha = \frac{1}{2},$$

where we applied Eqn. (5.17). Similarly, by taking the constants out of the integrations, the second integral in Eqn. (5.19) becomes,

$$\int_{-\infty}^{\infty} \frac{\partial h_\alpha\left(\phi_\alpha^0\right)}{\partial \phi_\alpha} \frac{\partial \phi_\alpha^0}{\partial \eta} \partial \eta \quad = \quad \int_{1}^{0} \frac{\partial h_\alpha\left(\phi_\alpha^0\right)}{\partial \phi_\alpha} \partial \phi_\alpha^0 = h_\alpha\left(0\right) - h\left(1\right) = -1.$$

With the evaluated integrals, the solvability condition (5.19) reads,

$$u^0 \quad = \quad \frac{\tilde{\tau} v}{g}.$$

Substituting the values for the parameters $\tilde{\tau}$ and $g$ in dimensional units reads,

$$(T - T_m^\alpha) \quad = \quad -\frac{\tau_{\alpha\beta} V T_m^\alpha}{L_\alpha},$$
$$= \quad -\overline{\beta}^0 V.$$

Comparing the terms we derived the zeroth order kinetic coefficient in terms of the phase-field parameters which is also the *sharp interface limit* given by,

$$\overline{\beta}^0 \quad = \quad \frac{\tau_{\alpha\beta} T_m^\alpha}{L_\alpha}.$$

Hence $\tau_{\alpha\beta}$ has the units $\dfrac{Js}{m^4}$. In case of using an entropy functional, $\overline{\beta}^0$ is a function of the temperature and hence can be written equivalently as,

$$\overline{\beta}^0 \quad = \quad \frac{\omega_{\alpha\beta} T T_m^\alpha}{L_\alpha},$$

where $\omega_{\alpha\beta}$ has the units $\dfrac{Js}{m^4 K}$.

In order to compute the *first order* correction to the kinetic coefficient, we write the $u$ equation at order $1/p$ which reads,

$$-v\frac{\partial u^0}{\partial \eta} \quad = \quad \frac{\partial^2 u^1}{\partial \eta^2} - v\frac{\partial h_\alpha\left(\phi_\alpha^0\right)}{\partial \eta}.$$

From the $u$ equation at order $1/p^2$ we have the result $u_0$ is constant. Hence, this does not contribute to the $u$ equation at order $1/p$. Integrating the resulting equation gives,

$$\frac{\partial u^1}{\partial \eta} \;=\; v h_\alpha \left(\phi_\alpha^0\right) + A. \tag{5.20}$$

Employing matching condition in Eqn.(5.15) we derive,

$$\lim_{\eta \to \pm\infty} \frac{\partial u^1}{\partial \eta} \;=\; \left.\frac{\partial \widetilde{u^0}}{\partial x}\right|^{\pm},$$
$$=\; A + v h_\alpha \left(\phi_\alpha^0 \pm\right).$$

which can be used to derive the macroscopic gradients of the temperature field computed from both the bulk sides,

$$\left.\frac{\partial \widetilde{u^0}}{\partial x}\right|^{+} \;=\; A, \tag{5.21}$$

$$\left.\frac{\partial \widetilde{u^0}}{\partial x}\right|^{-} \;=\; A + v. \tag{5.22}$$

Subtracting, we at once get the *Stefan condition* at the lowest order which is,

$$v \;=\; \left.\frac{\partial \widetilde{u^0}}{\partial x}\right|^{-} - \left.\frac{\partial \widetilde{u^0}}{\partial x}\right|^{+}.$$

Integrating Eqn.(5.20) we get the total inner solution at order $(1/p)$ as,

$$u_1 \;=\; \overline{u_1} + v \int_0^\eta h_\alpha \left(\phi_\alpha^0\right) \partial\eta + A\eta. \tag{5.23}$$

The integration constant $\overline{u_1}$ is computed by inserting the equation for $u_1$ into the phase-field evolution equation at order $p^2$ which reads,

$$-\tilde{\tau} v \frac{\partial \phi_\alpha^1}{\partial \eta} \;=\; \frac{\partial^2 \phi_\alpha^2}{\partial \eta^2} + \frac{16}{\pi^2}\phi_\alpha^2 - \frac{g u^1}{2}\frac{\partial h_\alpha (\phi_\alpha)}{\partial \phi_\alpha} - \frac{g u^0}{2}\phi_\alpha^1 \frac{\partial^2 h_\alpha \left(\phi_\alpha^0\right)}{\partial \phi_\alpha{}^2}.$$

Identifying the operator $L$ the equation can be written in short as,

$$L\phi_\alpha^2 \;=\; -\tilde{\tau}v\frac{\partial \phi_\alpha^1}{\partial \eta} + \frac{gu^1}{2}\frac{\partial h_\alpha\left(\phi_\alpha^0\right)}{\partial \phi_\alpha} + \frac{gu^0}{2}\phi_\alpha^1\frac{\partial^2 h_\alpha\left(\phi_\alpha^0\right)}{\partial \phi_\alpha^2}.$$

from which the solvability condition for a non-trivial $\phi_\alpha^0$ implies that the R.H.S must be orthogonal to $\partial_\eta\phi_\alpha^0$ giving,

$$-\tilde{\tau}v\int_{-\infty}^{\infty}\frac{\partial \phi_\alpha^1}{\partial \eta}\frac{\partial \phi_\alpha^0}{\partial \eta}\partial\eta + \int_{-\infty}^{\infty}\frac{gu^1}{2}\frac{\partial h_\alpha\left(\phi_\alpha^0\right)}{\partial \phi_\alpha}\frac{\partial \phi_\alpha^0}{\partial \eta}\partial\eta +$$

$$\int_{-\infty}^{\infty}\frac{gu^0}{2}\phi_\alpha^1\frac{\partial^2 h_\alpha\left(\phi_\alpha^0\right)}{\partial \phi_\alpha^2}\frac{\partial \phi_\alpha^0}{\partial \eta}\partial\eta \;=\; 0. \quad (5.24)$$

To simplify the preceding solvability condition we need to evaluate the nature of the integrals. For the first term, we make use of the fact that $\phi_\alpha^1$ satisfies the Eqn.(5.18),

$$L\phi_\alpha^1 \;=\; -\tilde{\tau}v\frac{\partial \phi_\alpha^0}{\partial \eta} + \frac{gu^0}{2}\frac{\partial h_\alpha\left(\phi_\alpha^0\right)}{\partial \phi_\alpha}. \quad (5.25)$$

The phase-field profile $\phi_\alpha^0$ is in the case of an obstacle type potential part of a *sinus* curve and hence is an odd-function, which implies its derivative $\dfrac{\partial \phi_\alpha^0}{\partial \eta}$ is even. Similarly, the interpolation function $h_\alpha\left(\phi_\alpha^0\right)$ is an odd function and hence, $\dfrac{\partial h_\alpha\left(\phi_\alpha^0\right)}{\partial \phi_\alpha}$ is an *even* function. To realize this, we use a property of the interpolation function which is the anti-symmetry with respect to $\eta = 0$ or at the position where $\phi_\alpha = 1/2$ yielding,

$$h_\alpha\left(\phi_\alpha\left(\eta\right)\right) - \frac{1}{2} \;=\; \frac{1}{2} - h_\alpha\left(\phi_\alpha\left(-\eta\right)\right).$$

Differentiating both sides with respect to $\eta$ and using the *even* property of $\dfrac{\partial \phi_\alpha^0}{\partial \eta}$ we derive,

$$\frac{\partial h_\alpha\left(\phi_\alpha\left(\eta\right)\right)}{\partial \phi_\alpha^0} \;=\; \frac{\partial h_\alpha\left(\phi_\alpha\left(-\eta\right)\right)}{\partial \phi_\alpha^0}$$

implying $\dfrac{\partial h_\alpha\left(\phi_\alpha\left(\eta\right)\right)}{\partial \phi_\alpha^0}$ is even. Conversely, differentiating again, we get

that the second derivative $\dfrac{\partial^2 h_\alpha\left(\phi_\alpha\left(\eta\right)\right)}{\partial \phi_\alpha^0{}^2}$ is odd. Using these properties, we

directly find that the R.H.S of Eqn. (5.25) is even. Combined with the fact

that the operator $L$ is of the form $\dfrac{\partial^2}{\partial \eta^2} + \dfrac{16}{\pi^2}$, which does not change the

characteristic properties of the R.H.S., we derive that $\phi_\alpha^1$ is even. Putting
all the arguments together, we directly see that only the second integral
in the solvability condition Eqn. (5.24) survives and simplifies to,

$$\int_{-\infty}^{\infty} \frac{gu^1}{2} \frac{\partial h_\alpha\left(\phi_\alpha^0\right)}{\partial \phi_\alpha} \frac{\partial \phi_\alpha^0}{\partial \eta} \partial \eta \;=\; 0.$$

Substituting $u^1$ from Eqn. (5.23) into the solvability condition gives,

$$\frac{\overline{gu^1}}{2} \int_{-\infty}^{\infty} \frac{\partial h_\alpha\left(\phi_\alpha^0\right)}{\partial \phi_\alpha^0} \frac{\partial \phi_\alpha^0}{\partial \eta} \partial \eta + \frac{g}{2} v \int_{-\infty}^{\infty} \left[\int_0^{\eta} h_\alpha\left(\phi_\alpha^0\right) \partial \eta\right] \frac{\partial h_\alpha\left(\phi_\alpha^0\right)}{\partial \phi_\alpha^0} \frac{\partial \phi_\alpha^0}{\partial \eta} \partial \eta +$$

$$\frac{g}{2} \int_{-\infty}^{\infty} A\eta \frac{\partial h_\alpha\left(\phi_\alpha^0\right)}{\partial \phi_\alpha^0} \frac{\partial \phi_\alpha^0}{\partial \eta} \partial \eta = 0.$$

In the above, the last integral vanishes because $\eta$ is an odd function, and
the other two functions in the integral are even, rendering the integral as
odd. Thus, the constant $\overline{u^1}$ can be written as,

$$\frac{\overline{gu^1}}{2}\left(h\left(0\right) - h\left(1\right)\right) + \frac{g}{2} v \int_{-\infty}^{\infty} \left[\int_0^{\eta} h_\alpha\left(\phi_\alpha^0\right) \partial \eta\right] \frac{\partial h_\alpha\left(\phi_\alpha^0\right)}{\partial \phi_\alpha^0} \frac{\partial \phi_\alpha^0}{\partial \eta} \partial \eta \;=\; 0.$$

We introduce $\tilde{M}$ such that,

$$\tilde{M} \;=\; \int_{-\infty}^{\infty} \left[\int_0^{\eta} h_\alpha\left(\phi_\alpha^0\right) \partial \eta\right] \frac{\partial h_\alpha\left(\phi_\alpha^0\right)}{\partial \phi_\alpha^0} \frac{\partial \phi_\alpha^0}{\partial \eta} \partial \eta$$

and hence,

$$\overline{u^1} \;=\; \tilde{M} v.$$

Substituting the expression for $\overline{u^1}$ in Eqn. (5.23) we obtain,

$$u^1 \;=\; \tilde{M}v + v \int_0^\eta h_\alpha\left(\phi_\alpha^0\right)\partial\eta + A\eta.$$

Using the condition in Eqn. (5.22), the constant $A$ can be written in two ways, which are constrained to be equal by the Stefan condition. Upon substitution, the two formulations can be realized as follows,

$$u^1 \;=\; \tilde{M}v + v \int_0^\eta h_\alpha\left(\phi_\alpha^0\right)\partial\eta + \left(\left.\frac{\widetilde{\partial u^0}}{\partial x}\right|^{+}\right)\eta \quad \text{alternatively} \quad (5.26)$$

$$u^1 \;=\; \tilde{M}v + v \int_0^\eta h_\alpha\left(\phi_\alpha^0\right)\partial\eta + \left(\left.\frac{\widetilde{\partial u^0}}{\partial x}\right|^{-}\right)\eta - v\eta$$

$$\;=\; \tilde{M}v + v \int_0^\eta \left(h_\alpha\left(\phi_\alpha^0\right) - 1\right)\partial\eta + \left(\left.\frac{\widetilde{\partial u^0}}{\partial x}\right|^{-}\right)\eta. \qquad (5.27)$$

Using the matching condition in Eqn. (5.12) gives

$$\widetilde{u^1}\big|^{\pm} = \lim_{\eta\to\pm\infty} u^1 - \eta\left.\frac{\widetilde{\partial u^0}}{\partial x}\right|^{\pm},$$

thereby, we derive the positive limit by including Eqn. (5.26) and the negative limit by Eqn. (5.27) for the derivatives $\dfrac{\widetilde{\partial u^0}}{\partial x}$ as,

$$\widetilde{u^1}\bigg|^{+} \;=\; \tilde{M}v + v \int_0^\infty h_\alpha\left(\phi_\alpha^0\right)\partial\eta$$

$$\widetilde{u^1}\bigg|^{-} \;=\; \tilde{M}v + v \int_0^{-\infty} \left(h_\alpha\left(\phi_\alpha^0\right) - 1\right)\partial\eta.$$

Due to the construction of the interpolation functions as anti-symmetric functions, one can verify that the two integrals $\int_0^\infty h_\alpha\left(\phi_\alpha^0\right)\partial\eta$ and $\int_0^{-\infty}\left(h_\alpha\left(\phi_\alpha^0\right) - 1\right)\partial\eta$ are equal and hence the macroscopic value of the outer solution at first order, $\widetilde{u^1}$ is unique from both sides. We denote the integral in the preceding equation as $\tilde{F}$ and hence $\widetilde{u^1}$ can be written as,

$$\widetilde{u^1} \;=\; v\left(\tilde{M} + \tilde{F}\right).$$

Putting the dimensions back, we have,

$$\widetilde{\Delta T^1} \;=\; v\frac{L_\alpha}{C_v}\left(\tilde{M} + \tilde{F}\right)$$

Combining the undercoolings at the different orders, we have the total macroscopic undercooling as,

$$\widetilde{\Delta T} = \widetilde{\Delta T^0} + p\widetilde{\Delta T^1},$$

which can equivalently be expressed in terms of the kinetic coefficients by,

$$-\widetilde{\bar{\beta}}V = -\widetilde{\bar{\beta}^0}V + p\widetilde{\bar{\beta}^1}V$$
$$= -\frac{\tau_{\alpha\beta}T_m}{L_\alpha}V + \varepsilon\frac{L_\alpha}{K}V\left(\tilde{M} + \tilde{F}\right).$$

Comparing terms, we have,

$$\widetilde{\bar{\beta}} \;=\; \widetilde{\bar{\beta}^0} - p\widetilde{\bar{\beta}^1}$$
$$= \;\frac{\tau_{\alpha\beta}T_m}{L_\alpha} - \varepsilon\frac{L_\alpha}{K}\left(\tilde{M} + \tilde{F}\right) \qquad (5.28)$$

For the commonly considered interpolation polynomials being the cubic and the quartic type polynomial, when used along with the obstacle potential, the values of $\tilde{F}$ and $\tilde{M}$ are tabulated below,

| | $\tilde{M}$ | $\tilde{F}$ |
|---|---|---|
| $h\left(\phi_\alpha\right) = \phi_\alpha^2\left(3 - 2\phi_\alpha\right)$ | 0.063828 | 0.158741 |
| $h\left(\phi_\alpha\right) = \phi_\alpha^3\left(10 - 15\phi_\alpha + 6\phi_\alpha^2\right)$ | 0.052935 | 0.129288 |

Eqn. (5.28) is the revolutionary result, first obtained by Karma in [48] and is known as the *thin interface limit*. It shows that there is the possibility to choose parameters $\tau_{\alpha\beta}$ and interface width parameter $\varepsilon$ in a way that the effective $\overline{\beta}$ vanishes and hence simulations at the vanishing interface kinetics limit can be performed. This can be achieved by manipulating Eqn. (5.28) as,

$$\tau_{\alpha\beta} = \varepsilon \frac{L_\alpha^2}{T_m K} \left( \tilde{M} + \tilde{F} \right). \tag{5.29}$$

In order to achieve low interface kinetics using the sharp interface limit, the value of $\tau_{\alpha\beta}$ becomes prohibitively lower and hence the time step reduces to a computationally unfeasible value. However with the thin interface limit, one has the benefits of an increased interface width that can be chosen for the simulation which can recover the sharp interface free boundary problem along with the condition that the time scales for the simulation can be enhanced for problems with vanishing interface kinetics.

## 5.4. Benchmarks

In order to benchmark our calculations, we choose the Ni system for consideration. The system parameters involve the latent heat of the system which is given by $L_\alpha = 0.30$, which is non-dimensionalized using energy scale as, $C_v^s T_m = 1.52 \times 10^6 \, J/m^3$, where $T_m = 1748K$ is the melting point of the solid and $C_v^s$ is the specific heat of Ni. The surface entropy density of the solid-liquid interface is given by $\gamma_{\alpha\beta} = 0.167 \times 10^{-3} \, J/m^2 K$. We perform several simulations of the growth of a planar front at a given non-dimensional temperature given by $T = 0.96$ $(T_m = 1.0)$, with variation in the interface widths, affected by changing the parameter $\varepsilon$. The results plotted in Fig.5.1 show the invariance of the front velocity upon change in $\varepsilon$. To achieve this, we affect $\overline{\beta} = 0$, through an appropriate choice of the parameter $\tau_{\alpha\beta}$, given by the expression in Eqn.(5.29). The existence of such a range in $\varepsilon$, where the velocity is relatively constant, denotes that we have been able to effectively eliminate the first order contributions of the interface width to the interface kinetic coefficient $\overline{\beta}$. The specific heats and the thermal conductivities of the liquid are close to the solid and hence the validity of our assumptions used in the calculations remain.

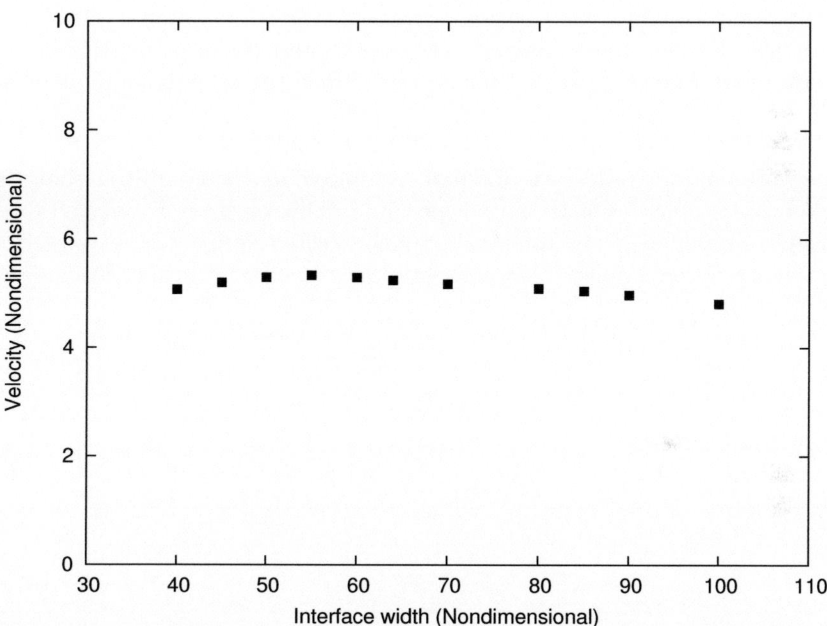

Figure 5.1.: Velocity of a pure nickel planar front simulated with different $\varepsilon$ at a fixed nondimensional temperature of T=0.96.

## 5.5. Concluding remarks

In the present work, we present the thin-interface asymptotics of pure material solidification for the case of the double obstacle potential. This work extends the computational simplicity of the obstacle potential to the regime of quantitative simulations, where the results are invariant upon change in the interface widths. Further, the derivations enable the adjustment of simulation parameters for performing vanishing interface kinetics. In the above analysis, we have assumed the properties of the thermal conductivity $K$ or the thermal diffusivity as independent of $\phi_\alpha$ which renders the two limits $\widetilde{u^1}\Big|^+$ and $\widetilde{u^1}\Big|^-$ equal. In the event that the thermal conductivities are unequal, we arrive at the condition that the two macroscopic limits are no longer equal. This property is referred to as "heat trapping" e.g. [6]. Removal of such effects requires the introduction of an expression similar to the anti-trapping term for solute diffusion, [46]. While this is derived for the case of one sided diffusivities, the case of non-vanishing thermal diffusivity needs a treatment similar as in [83].

# Chapter 6

# Grand potential formulation and asymptotics

# 6.1. Introduction and model modification

Phase-field modeling has been used for alloy solidification for about a decade and the principle ideas are fairly well known. However, it is necessary to highlight the importance of certain modifications without which large scale quantitative microstructure simulations of the order of micrometers are not possible. In the present paper we describe a modification of the multi-phase field model described in [33]. The foundation of this particular model is the entropy functional written as follows,

$$\mathcal{S}\left(e, \boldsymbol{c}, \boldsymbol{\phi}\right) \;\; = \;\; \int_{\Omega} \left( s\left(e, \boldsymbol{c}, \boldsymbol{\phi}\right) - \left( \varepsilon a\left(\boldsymbol{\phi}, \nabla\boldsymbol{\phi}\right) + \frac{1}{\varepsilon} w\left(\boldsymbol{\phi}\right) \right) \right) d\Omega,$$

where $e$ is the internal energy of the system, and $s$ is the bulk entropy density and $w$ is the surface potential of the system. $\boldsymbol{c} = \left(c_1 \ldots c_K\right)$, is vector with the compositions of the $K$ components and $\boldsymbol{\phi} = \left(\phi_1 \ldots \phi_N\right)$ are the volume fractions of the $N$ phases in the system. An equivalent form can be defined as a free energy functional at a given temperature $T$,

$$\mathcal{F}\left(T, \boldsymbol{c}, \boldsymbol{\phi}\right) \;\; = \;\; \int_{\Omega} \left( f\left(T, \boldsymbol{c}, \boldsymbol{\phi}\right) + \left( \varepsilon \tilde{a}\left(\boldsymbol{\phi}, \nabla\boldsymbol{\phi}\right) + \frac{1}{\varepsilon} \tilde{w}\left(\boldsymbol{\phi}\right) \right) \right) d\Omega.$$

The uniqueness lies in the usage of the double obstacle potential $w\left(\boldsymbol{\phi}\right) = \{\sum_{\alpha<\beta}^{N,N} \gamma_{\alpha\beta} \frac{16}{\pi^2} \phi_\alpha \phi_\beta$, when $\left(\phi_\alpha, \phi_\beta > 0 \text{ and } \phi_\alpha + \phi_\beta = 1\right)$ and $\infty$ elsewhere$\}$ in describing the surface entropy potential where the relations $\tilde{a}\left(\boldsymbol{\phi}, \nabla\boldsymbol{\phi}\right) = T a\left(\boldsymbol{\phi}, \nabla\boldsymbol{\phi}\right)$ and $\tilde{w}\left(\boldsymbol{\phi}\right) = T w\left(\boldsymbol{\phi}\right)$ are valid. However, if the free energies are interpolated as in [33, 79], $f = \sum_{\alpha=1}^{N} f_\alpha\left(T, \boldsymbol{c}, \boldsymbol{\phi}\right) h_\alpha\left(\boldsymbol{\phi}\right)$, where $f_\alpha\left(T, \boldsymbol{c}\right)$ is the bulk free energy density of phase $\alpha$, and $h_\alpha\left(\boldsymbol{\phi}\right)$ is an interpolation function for the phase $\alpha$, two problems exist,
i) The surface energy $\tilde{\sigma}_{\alpha\beta}$ of an $\alpha\beta$ interface is a function of the chemical free energy density landscape in the system and
ii) the equilibrium interface width $\widetilde{\Lambda}_{\alpha\beta}$ becomes far too restrictive for simulating large scale microstructures.
These restrictions of the model will be highlighted in more detail in the following discussion. The equilibrium equation in 1D for two phases, $\alpha$ and

$\beta$, where $\phi_\alpha + \phi_\beta = 1$, starting from the interpolation of the free energies can be written as follows,

$$\gamma_{\alpha\beta}\varepsilon \frac{\partial^2 \phi_\alpha}{\partial x^2} = -\frac{16}{\pi^2}\frac{\gamma_{\alpha\beta}}{2\varepsilon}(1 - 2\phi_\alpha) - \frac{1}{2T}\frac{df}{d\phi_\alpha} + \frac{1}{2T}\sum_{i=1}^{K-1}\mu_i \frac{dc_i}{d\phi_\alpha} \quad (6.1)$$

$$= -\frac{16}{\pi^2}\frac{\gamma_{\alpha\beta}}{2\varepsilon}(1 - 2\phi_\alpha) - \frac{1}{2T}\frac{d}{d\phi_\alpha}\left(f - \sum_{i=1}^{K-1}\mu_i c_i\right), \quad (6.2)$$

where $\boldsymbol{\mu} = (\mu_i \ldots \mu_{K-1})$ is the vector consisting of the $K-1$ equilibrium chemical potentials of the system at the given system temperature. Eqn. (6.2) can be used to derive the stationary solution of the phase-field $\phi_\alpha$, which can be used to derive expressions for the surface energy of a binary interface, $\tilde{\sigma}_{\alpha\beta}$ and equilibrium interface width $\tilde{\Lambda}_{\alpha\beta}$ as,

$$\tilde{\sigma}_{\alpha\beta} = 2T\gamma_{\alpha\beta}\int_0^1 \sqrt{\left(\frac{16}{\pi^2}\phi_\alpha(1 - \phi_\alpha) + \frac{\varepsilon}{\gamma_{\alpha\beta}T}(\Delta\Psi(T, \boldsymbol{c}, \phi_\alpha))\right)}d\phi_\alpha \quad (6.3)$$

$$\tilde{\Lambda}_{\alpha\beta} = \varepsilon \int_0^1 \frac{d\phi_\alpha}{\sqrt{\left(\frac{16}{\pi^2}\phi_\alpha(1 - \phi_\alpha) + \frac{\varepsilon}{\gamma_{\alpha\beta}T}(\Delta\Psi(T, \boldsymbol{c}, \phi_\alpha))\right)}}. \quad (6.4)$$

Here $\gamma_{\alpha\beta}$ is a term in the surface entropy density, $\varepsilon$ is a factor related to the length scale of the interface and $\Delta\Psi(T, \boldsymbol{c}, \phi_\alpha) = \left(f - \sum_{i=1}^{K-1}\mu_i c_i\right) - \left(f - \sum_{i=1}^{K-1}\mu_i c_i\right)_{\phi_\alpha=0}$ is the grand chemical potential difference between values at the interface and that of the bulk phases in equilibrium. At equilibrium, the terms $\left(f - \sum_{i=1}^{K-1}\mu_i c_i\right)_{\phi_\alpha=0}$ and $\left(f - \sum_{i=1}^{k-1}\mu_i c_i\right)_{\phi_\alpha=1}$ are equal and Fig. 6.1 plots the variation of the term $\Delta\Psi(T, \boldsymbol{c}, \phi_\alpha)$. We clearly see that the term $\frac{1}{2}\frac{d}{d\phi_\alpha}\left(f - \sum_{i=1}^{K-1}\mu_i c_i\right)$ is non-zero across the interface. The area under the curve is the grand chemical potential excess at the interface. This contribution affects the equilibrium shape and properties of the interface. From the above, it is evident that,

- The parameters $\tilde{\sigma}_{\alpha\beta}$ and $\tilde{\Lambda}_{\alpha\beta}$ cannot be fixed independently of the grand chemical potential contribution in the form $\Delta\Psi(T, \boldsymbol{c}, \phi_\alpha)$.

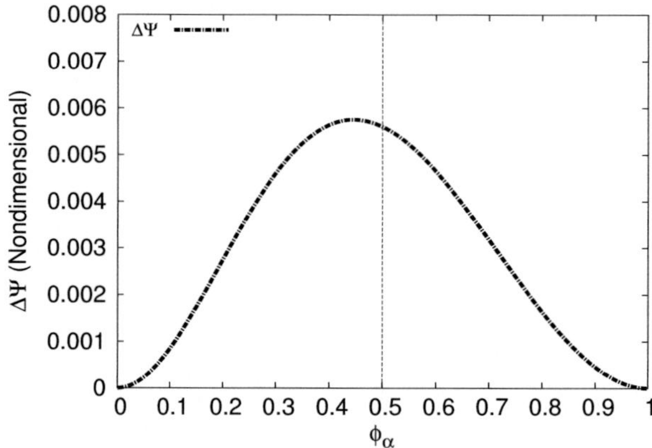

**Figure 6.1.:** The grandchemical potential difference varies across the interface and has a form similar to that of a potential. At equilibrium, the two phases are at the same grand chemical potential which is seen qualitatively from the graph. Notice also the asymmetry of the potential around $\phi_\alpha = 0.5$, which is inherited from the asymmetry in the chemical free energy states of the two phases.

- Given the required $\tilde{\sigma}_{\alpha\beta}$ and $\tilde{\Lambda}_{\alpha\beta}$ the simulation parameters $\gamma_{\alpha\beta}$ and $\varepsilon$ can be determined by simultaneously solving the Eqns. (6.3) and (6.4). Notice, that even though we have one parameter $\varepsilon$, the resulting interface thicknesses can be different, depending on the excess $\Delta\Psi\left(T, \boldsymbol{c}, \phi_\alpha\right)$.

For very large chemical excess contributions in the form of $\Delta\Psi\left(T, \boldsymbol{c}, \phi_\alpha\right)$, the Eqns. (6.3) and (6.4) can be written approximately as,

$$
\tilde{\sigma}_{\alpha\beta} = 2\sqrt{T\gamma_{\alpha\beta}\varepsilon}\int_0^1 \sqrt{(\Delta\Psi\left(T, \boldsymbol{c}, \phi_\alpha\right))}d\phi_\alpha
$$

$$
\tilde{\Lambda}_{\alpha\beta} = \sqrt{T\gamma_{\alpha\beta}\varepsilon}\int_0^1 \frac{d\phi_\alpha}{\sqrt{(\Delta\Psi\left(T, \boldsymbol{c}, \phi_\alpha\right))}}.
$$

In this case $\tilde{\sigma}_{\alpha\beta}$ and $\tilde{\Lambda}_{\alpha\beta}$ are no longer independent. The term $\dfrac{\tilde{\sigma}_{\alpha\beta}}{\tilde{\Lambda}_{\alpha\beta}}$ becomes just a function of the chemical free energy of the system and independent of the terms $\gamma_{\alpha\beta}$ and $\varepsilon$. This implies that once a value for $\tilde{\sigma}_{\alpha\beta}$ is chosen, the value of $\tilde{\Lambda}_{\alpha\beta}$ is fixed, and for certain choices of $\tilde{\sigma}_{\alpha\beta}$, the $\tilde{\Lambda}_{\alpha\beta}$ gets prohibitively lower, which makes simulation of larger domain structures unfeasible. These relationships have been studied fairly extensively in the past decade, and two principle solutions have been suggested [27, 30, 55, 114]. The ideology is to completely avoid any contribution of the grand chemical potential excess contribution to the interface excess. This implies, the stationary solution is independent of any chemical contribution. While this is achieved in the work by [27, 55, 114] through the use of different concentration fields $c_i^\alpha$ in each phase, the same is affected for dilute alloys, with a single concentration field but through the use of effective interpolation functions to interpolate the entropy and enthalpy contributions to the free energy. The common idea is that the driving force for phase transformation is the grand potential difference between the phases at the same chemical potential. We motivate a similar idea from the following discussion.

## 6.1.1. Motivation

Consider the phase-field evolution equation in 1D at the lowest order in $\varepsilon$, which is a parameter related to the interface thickness:

$$\omega_{\alpha\beta}\varepsilon\frac{\partial\phi_\alpha}{\partial t} = \gamma_{\alpha\beta}\varepsilon\frac{\partial^2\phi_\alpha}{\partial x^2} - \frac{16}{\pi^2}\frac{\gamma_{\alpha\beta}}{2\varepsilon}(1-2\phi_\alpha) - \frac{1}{2T}\frac{d}{d\phi_\alpha}\left(f - \sum_{i=1}^{K-1}\mu_i c_i\right),$$

where $\tau_{\alpha\beta}$ is the relaxation constant of the interface. This is also the evolution equation at the *sharp-interface limit* for this model [134]. The chemical potential $\boldsymbol{\mu} = (\mu_1 \ldots \mu_{K-1})$ is constant across the interface in this limit. For small velocities, the evolution equation in moving co-ordinate frame in 1D, at steady state velocity V reads,

$$-V\omega_{\alpha\beta}\varepsilon\frac{d\phi_\alpha}{dx} = \gamma_{\alpha\beta}\varepsilon\frac{d^2\phi_\alpha}{dx^2} - \frac{16}{\pi^2}\frac{\gamma_{\alpha\beta}}{2\varepsilon}(1-2\phi_\alpha) - \frac{1}{2T}\frac{d}{d\phi_\alpha}\left(f - \sum_{i=1}^{K-1}\mu_i c_i\right).$$

*It is important to note that the moving frame is moving with velocity V along with the interface which is denoted by the contour line $\phi_\alpha = 0.5$.* Multiplying with $\dfrac{d\phi_\alpha}{dx}$ on both sides and integrating we get,

$$-V\omega_{\alpha\beta}\varepsilon\int_{-\infty}^{\infty}\left(\frac{d\phi_\alpha}{dx}\right)^2 dx = \int_{-\infty}^{\infty}\gamma_{\alpha\beta}\varepsilon\frac{d^2\phi_\alpha}{dx^2}\frac{d\phi_\alpha}{dx}dx -$$

$$\int_{-\infty}^{\infty}\frac{16}{\pi^2}\frac{\gamma_{\alpha\beta}}{2\varepsilon}(1-2\phi_\alpha)\frac{d\phi_\alpha}{dx}dx - \int_{-\infty}^{\infty}\frac{1}{2T}\frac{d}{d\phi_\alpha}\left(f - \sum_{i=1}^{K-1}\mu_i c_i\right)\frac{d\phi_\alpha}{dx}dx.$$

We denote the integral $\int_{-\infty}^{\infty}\left(\dfrac{d\phi_\alpha}{dx}\right)^2 dx$ as (I) and elaborate the other integrals as follows;

$$-V\omega_{\alpha\beta}\varepsilon\mathrm{I} = \frac{\gamma_{\alpha\beta}\varepsilon}{2}\left(\frac{d\phi_\alpha}{dx}\right)^2\Big|_{-\infty}^{\infty} - \frac{16}{\pi^2}\frac{\gamma_{\alpha\beta}}{2\varepsilon}\phi_\alpha(1-\phi_\alpha)\Big|_0^1 -$$

$$\frac{1}{2T}\left(f - \sum_{i=1}^{K-1}\mu_i c_i\right)\Big|_0^1.$$

The first two integrals on the right hand side drop out to zero and so the velocity of the interface can be written as,

$$- V\omega_{\alpha\beta}\varepsilon\mathrm{I} \;=\; \frac{1}{2T}\left(f - \sum_{i=1}^{K-1}\mu_i c_i\right)\Bigg|_0^1, \tag{6.5}$$

Clearly, the interface mobility is proportional to the difference of the *grand chemical potential* of the two phases. Adequately, the driving force for phase transformation in alloys in the sharp interface limit is the difference of the grand potentials of the two bulk phases. The evolution equations drive the system in a direction to reduce the difference of grand potentials between the bulk phases. This being the case, the motivation arises to formulate the phase-field model in terms of a grand potential functional for the case of alloys.

## 6.1.2. Model modification

We write the grand potential density $\Psi$, as an interpolation of the individual grand potential densities $\Psi_\alpha$, where $\Psi_\alpha$ are functions of the chemical potential $\boldsymbol{\mu}$ and temperature T in the system,

$$\Psi\left(T,\boldsymbol{\mu},\boldsymbol{\phi}\right) \;=\; \sum_{\alpha=1}^{N}\Psi_\alpha\left(T,\boldsymbol{\mu}\right)h_\alpha\left(\boldsymbol{\phi}\right) \quad \text{with,} \tag{6.6}$$

$$\Psi_\alpha\left(T,\boldsymbol{\mu}\right) \;=\; f_\alpha\left(\boldsymbol{c}^\alpha\left(\boldsymbol{\mu}\right),T\right) - \sum_{i=1}^{K-1}\mu_i c_i^\alpha\left(\boldsymbol{\mu},T\right).$$

The concentration $c_i^\alpha\left(\boldsymbol{\mu},T\right)$ is an inverse of the function $\mu_i^\alpha\left(\boldsymbol{c},T\right)$ for every phase $\alpha$ and component $i$. From Eqn.(6.6) the following relation can be derived,

$$\frac{\partial\Psi\left(T,\boldsymbol{\mu},\boldsymbol{\phi}\right)}{\partial\mu_i} \;=\; \sum_{\alpha=1}^{N}\frac{\partial\Psi_\alpha\left(T,\boldsymbol{\mu}\right)}{\partial\mu_i}h_\alpha\left(\boldsymbol{\phi}\right).$$

Since, the grand potential density $\Psi\left(T,\boldsymbol{\mu},\boldsymbol{\phi}\right)$, is the *Legendre transform* of the free energy density of the system $f\left(T,\boldsymbol{c},\boldsymbol{\phi}\right)$, and from their coupled relation $\dfrac{\partial\Psi\left(T,\boldsymbol{\mu},\boldsymbol{\phi}\right)}{\partial\mu_i}=-c_i$, it follows that,

$$c_i \;=\; \sum_{\alpha=1}^{N} c_i^{\alpha}\left(\boldsymbol{\mu},T\right)h_{\alpha}\left(\boldsymbol{\phi}\right). \tag{6.7}$$

The above is the constraint used in [27, 55, 114] to determine the concentrations $c_i^{\alpha}$ in the interface along with the condition that the phase concentrations $c_i^{\alpha}$ are related by the condition of common equilibrium chemical potential among all the phases. This however derives elegantly starting from the grand potential functional. It is important to note that the entire structure rests on the invertibility of the function $\mu^{\alpha}\left(\boldsymbol{c},T\right)$. This would result in a unique grand potential for a given $\boldsymbol{\mu}$.

Since at equilibrium the grand potential of the phases are equal, for a two phase interface we can write,

$$\Psi\left(T,\boldsymbol{\mu}_{eq}\right) \;=\; \Psi_{\alpha}\left(T,\boldsymbol{\mu}_{eq}\right)=\Psi_{\beta}\left(T,\boldsymbol{\mu}_{eq}\right).$$

This implies that at equilibrium the surface energy has no contribution from the chemical free energy, since the grand chemical potential excess $\Delta\Psi$ is zero. The consequence of this is that the surface energy $\tilde{\sigma}_{\alpha\beta}$ is the same as the simulation parameter $\gamma_{\alpha\beta}T$. Also, it can be derived that the equilibrium interface width $\tilde{\Lambda}_{\alpha\beta}$ is independent of the chemical free energy of the system and is related to constant $\varepsilon$ by the relation $\dfrac{\pi^2}{4}\varepsilon$ for the obstacle potential. The grand chemical potential difference can be visualized as in Figure 6.2. A corollary of the above discussion is that the free energy of a mixture of two phases for alloys is not the interpolation of the free energies of the respective phases at a given concentration but it is a mixture of the phases at the respective concentrations at which they are at thermodynamic equilibrium i.e. at the same chemical potential $\boldsymbol{\mu}$. This

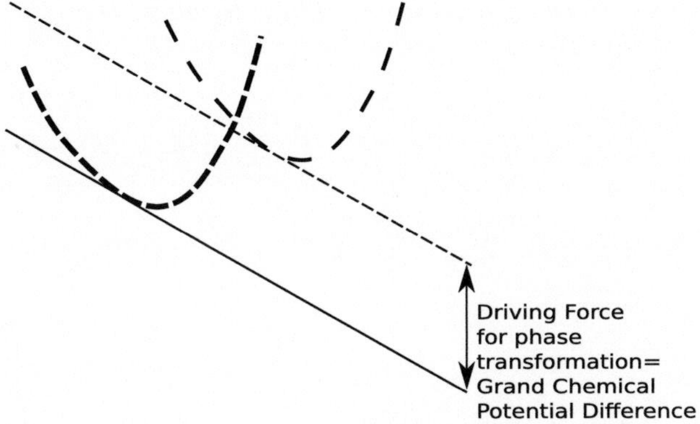


**Figure 6.2.:** Illustration of the driving force for phase transformation between two phases.

can be realized through the reverse Legendre transform of the expression in Eqn. (6.6), which gives,

$$f\left(T, \boldsymbol{c}, \boldsymbol{\phi}\right) \;=\; \sum_{\alpha=1}^{N} f_\alpha\left(\boldsymbol{c}^\alpha\left(\boldsymbol{\mu}, T\right), T\right) h_\alpha\left(\phi_\alpha\right).$$

This is the start point of the derivation of the KKS(Kim,Kim, Suzuki) model [55]. In summary, the principal result is that we write the evolution equations using the chemical potential $\boldsymbol{\mu}$ which is analogous to $T$ for the case of pure materials. The driving force, which is the difference of free energies in the case of pure materials translates to the difference of grand potentials for alloys. *Note: Strictly speaking, the grand potential is defined in terms of the number of particles of the various components written as $F - \sum_{i=1}^{K-1} \mu_i N_i$, where $F$ is the free energy of the system of $N$ particles, and $N_i$ is the number of particles of component $i$, while $\mu_i = \dfrac{\partial F}{\partial N_i}$. In the discussion on phase-field we require the energy densities of the respective phases, and hence, the energy of the system is generally divided by the volume of the system, which for the case of 1 mole of particles would be $V_m$, which is the molar volume. Also, the number of particles can be* *written in terms of the concentrations "mole fraction" through the relation $N_i = c_i N_o$, where $N_o$ is the Avogadro number. Utilizing this it is easy to see that $N_i \dfrac{\partial F}{\partial N_i} = V_m c_i \dfrac{\partial f}{\partial c_i}$, where $F = fV_m$ and we have assumed the molar volumes of all particles the same. This implies that the total grand potential can be written as $V_m \left( f - \sum_{i=1}^{K-1} \mu_i c_i \right)$, giving us the grand potential density as $\left( f - \sum_{i=1}^{K-1} \mu_i c_i \right)$. This is the form, which is used in the entire dissertation.*

### 6.1.3. Evolution equations

The evolution equations for the phase and concentration fields can be evaluated in the standard way. Phase evolution is determined by the phenomenological minimization of the modified functional which is formulated as the *grand potential functional*,

$$\Omega\left(T, \boldsymbol{\mu}, \boldsymbol{\phi}\right) = \int_{\Omega} \left( \Psi\left(T, \boldsymbol{\mu}, \boldsymbol{\phi}\right) + \left( \varepsilon \tilde{a}\left(\boldsymbol{\phi}, \nabla\boldsymbol{\phi}\right) + \frac{1}{\varepsilon}\tilde{w}\left(\boldsymbol{\phi}\right) \right) \right) d\Omega.$$

The concentration fields are obtained by a mass conservation equation for each of the $K-1$ independent concentration variables $c_i$. The evolution equation for the N phase-field variables can be written as,

$$\tau\varepsilon\frac{\partial\phi_\alpha}{\partial t} = \varepsilon\left( \nabla\cdot\frac{\partial\tilde{a}\left(\boldsymbol{\phi}, \nabla\boldsymbol{\phi}\right)}{\partial\nabla\phi_\alpha} - \frac{\partial\tilde{a}\left(\boldsymbol{\phi}, \nabla\boldsymbol{\phi}\right)}{\partial\phi_\alpha} \right) - \frac{1}{\varepsilon}\frac{\partial\tilde{w}\left(\boldsymbol{\phi}\right)}{\partial\phi_\alpha} -$$
$$\frac{\partial\Psi\left(T, \boldsymbol{\mu}, \boldsymbol{\phi}\right)}{\partial\phi_\alpha} - \Lambda,$$

where $\Lambda$ is the Lagrange parameter to maintain the constraint $\sum_{\alpha=1}^{N}\phi_\alpha = 1$. $a\left(\boldsymbol{\phi}, \nabla\boldsymbol{\phi}\right)$ represents the gradient energy density and has the form,

$$\tilde{a}\left(\boldsymbol{\phi}, \nabla\boldsymbol{\phi}\right) = \sum_{\substack{\alpha,\beta=1 \\ (\alpha<\beta)}}^{N,N} \tilde{\sigma}_{\alpha\beta}\left[a_c\left(q_{\alpha\beta}\right)\right]^2 |q_{\alpha\beta}|^2,$$

where $q_{\alpha\beta} = (\phi_\alpha \nabla \phi_\beta - \phi_\beta \nabla \phi_\alpha)$ is a normal vector to the $\alpha\beta$ interface. $a_c(q_{\alpha\beta})$ describes the form of the anisotropy of the evolving phase boundary. The double obstacle potential $\tilde{w}(\phi)$ which was also previously described in [33, 79] can be written as,

$$\tilde{w}(\phi) = \frac{16}{\pi^2} \sum_{\substack{\alpha,\beta=1 \\ (\alpha<\beta)}}^{N,N} \tilde{\sigma}_{\alpha\beta} \phi_\alpha \phi_\beta,$$

where $\tilde{\sigma}_{\alpha\beta}$ is the surface energy . The parameter $\tau$ is written as $\dfrac{\sum_{\alpha<\beta}^{N,N} \tau_{\alpha\beta} \phi_\alpha \phi_\beta}{\sum_{\alpha<\beta}^{N,N} \phi_\alpha \phi_\beta}$, where $\tau_{\alpha\beta}$ is the relaxation constant of the $\alpha\beta$ interface. The evolution equation for the concentration fields can be derived as,

$$\frac{\partial c_i}{\partial t} = \nabla \cdot \left( \sum_{j=1}^{K-1} M_{ij}(\phi) \nabla \mu_j \right). \tag{6.8}$$

Here, $M_{ij}(\phi)$ is the mobility of the interface, where the individual phase mobilities are interpolated as,

$$M_{ij}(\phi) = \sum_{\alpha=1}^{N-1} M_{ij}^\alpha g_\alpha(\phi),$$

where each of the $M_{ij}^\alpha$ is defined using the expression,

$$M_{ij}^\alpha = D_{ij}^\alpha \frac{\partial c_i^\alpha(\mu, T)}{\partial \mu_j}.$$

The function $g_\alpha(\phi)$ interpolates the mobilities and is in general not same as $h_\alpha(\phi)$ which interpolates the grand potentials. $D_{ij}^\alpha$ are the interdiffusivities in each phase $\alpha$. Both the evolution equations require the information about the chemical potential $\mu$. Two possibilities exist to determine the unknown chemical potential $\mu$.

- The chemical potential $\mu$ can be derived from the constraint relation (6.7). The $K-1$ independent components $\mu_i$ are determined by simultaneously solving the $K-1$ constraints for each of the $K-1$

independent concentration variables $c_i$, from the given values of $c_i$ and $\phi_\alpha$ at a given grid point. A Newton iteration scheme can be used for solving the system of equations,

$$\left\{\mu_i^{n+1}\right\} = \left\{\mu_i^n\right\} - \left[\sum_{\alpha=1}^{N} h_\alpha\left(\phi\right)\frac{\partial c_i^\alpha\left(\boldsymbol{\mu}^n,T\right)}{\partial\mu_j}\right]_{ij}^{-1} \times$$

$$\left\{c_i - \sum_{\alpha=1}^{N} c_i^\alpha\left(\boldsymbol{\mu}^n,T\right)h_\alpha\left(\phi\right)\right\}, \tag{6.9}$$

where $\{\}$ represents a vector while $[]$ denotes a matrix. This is precisely the approach in the KKS model [55]. However, there is a substantial difference, in that we propose to solve directly for the thermodynamic variable $\boldsymbol{\mu}$, which relate the phase concentrations $c_i^\alpha$, instead of solving for phase concentrations themselves. This is possible because the concentrations $c_i^\alpha\left(\boldsymbol{\mu},T\right)$ are written as explicit functions of the thermodynamic variable $\boldsymbol{\mu}$. The method also bears similarity to the method of [27, 114], where a partition relation is used to close the relationship between the phase concentrations.

- Alternatively explicit evolution equations for all the $K-1$ independent chemical potentials, can be formulated by inserting the constraint equation (6.7) into the evolution equation for the concentration field, Eqn. (6.8). For a two phase binary alloy i.e. $(\phi_\alpha + \phi_\beta = 1)$ and $c_A + c_B = 1$, the evolution equation can be written down as follows,

$$\left(\frac{\partial c^\alpha\left(\mu,T\right)}{\partial\mu}h_\alpha\left(\phi\right) + \frac{\partial c^\beta\left(\mu,T\right)}{\partial\mu}\left(1 - h_\alpha\left(\phi\right)\right)\right)\frac{\partial\mu}{\partial t} =$$

$$\nabla\cdot\left(\left(D^\alpha g_\alpha\left(\phi\right)\frac{\partial c^\alpha\left(\mu,T\right)}{\partial\mu} + D^\beta\left(1 - g_\alpha\left(\phi\right)\right)\frac{\partial c^\beta\left(\mu,T\right)}{\partial\mu}\right)\nabla\mu\right) -$$

$$\left(c^\alpha\left(\mu,T\right) - c^\beta\left(\mu,T\right)\right)\frac{\partial h_\alpha\left(\phi\right)}{\partial t},$$

where $c^{\alpha,\beta}\left(\mu\right)$ are the phase concentrations as functions of the independent chemical potential $\mu$. $D^\alpha$, $D^\beta$ are the independent interdiffusivities in the two respective phases. It is noteworthy that this equation looks very similar to the evolution equation of the temperature field in pure materials. The last term on the right hand side $c^\alpha\left(\mu, T\right) - c^\beta\left(\mu, T\right)$ corresponds to a source term for rejection of mass at the interface during growth, which is analogous to the release of latent heat in pure material solidification. For a general, multi-phase, multi-component system, the evolution equations for the components of the chemical potential $\boldsymbol{\mu}$ can be written in matrix form by,

$$\left\{\frac{\partial \mu_i}{\partial t}\right\} = \left[\sum_{\alpha=1}^{N} h_\alpha\left(\boldsymbol{\phi}\right) \frac{\partial c_i^\alpha\left(\boldsymbol{\mu}, T\right)}{\partial \mu_j}\right]_{ij}^{-1} \times$$
$$\left\{\nabla \cdot \sum_{j=1}^{K-1} M_{ij}\left(\boldsymbol{\phi}\right) \nabla \mu_j - \sum_\alpha c_i^\alpha\left(\boldsymbol{\mu}, T\right) \frac{\partial h_\alpha\left(\boldsymbol{\phi}\right)}{\partial t}\right\}. \quad (6.10)$$

The above derivation bears a lot of resemblance to the recent derivation by M.Plapp [91]. It is worth to comment on how the two methods compare in the computational complexity. For this, it is first essential to identify the similarity of the approaches, as can be seen by comparing Eqn. (6.9) and (6.10) as follows. Consider the case when Eqn. 6.9 is written for the case of binary which reads,

$$\mu^{n+1} = \mu^n + \frac{c - \sum_{\alpha=1}^{N} c^\alpha\left(\mu^n, T\right) h_\alpha\left(\boldsymbol{\phi}\right)}{\sum_{\alpha=1}^{N} \dfrac{\partial c^\alpha\left(\mu^n, T\right)}{\partial \mu} h_\alpha\left(\boldsymbol{\phi}\right)}, \quad (6.11)$$

where we intend to calculate the $\mu$ for the next time step $(t+1)$, $\mu^n$ being the start guess for the iteration which satisfies the equation $c - \sum_{\alpha=1}^{N} c^\alpha\left(\mu^n, T\right) h_\alpha\left(\boldsymbol{\phi}\right) = 0$ for the values of $c = c^o$ and $\phi = \phi^o$ at the time step $t$. Expanding the second term on the R.H.S in the Eqn. 6.11 for small change in time $\delta t$, we can write,

$$\mu^{n+1} - \mu^n = \frac{c^o - \sum_{\alpha=1}^{N} c^\alpha\left(\mu^n, T\right) h_\alpha\left(\phi^o\right)}{\sum_{\alpha=1}^{N} \dfrac{\partial c^\alpha\left(\mu^n, T\right)}{\partial \mu} h_\alpha\left(\phi^o\right)} +$$

$$\delta t \left( \frac{\dfrac{\partial c}{\partial t} - \sum_{\alpha=1}^{N} c^{\alpha}\left(\mu^{n}, T\right) \dfrac{\partial h_{\alpha}\left(\phi^{o}\right)}{\partial t}}{\sum_{\alpha=1}^{N} \dfrac{\partial c^{\alpha}\left(\mu^{n}, T\right)}{\partial \mu} h_{\alpha}\left(\phi^{o}\right)} \right) + O\left(\delta t^{2}\right).$$

Note, additional terms arise out of the linear expansion, but we simplify using the fact that $c^{0} - \sum_{\alpha=1}^{N} c^{\alpha}\left(\mu^{n}, T\right) h_{\alpha}\left(\phi^{0}\right) = 0$. Using the same fact, the preceding equation simplifies to,

$$\frac{\mu^{n+1} - \mu^{n}}{\delta t} = \left( \frac{\dfrac{\partial c}{\partial t} - \sum_{\alpha=1}^{N} c^{\alpha}\left(\mu^{n}, T\right) \dfrac{\partial h_{\alpha}\left(\phi^{o}\right)}{\partial t}}{\sum_{\alpha=1}^{N} \dfrac{\partial c^{\alpha}\left(\mu^{n}, T\right)}{\partial \mu} h_{\alpha}\left(\phi^{o}\right)} \right) + O\left(\delta t\right),$$

which in the region of small enough $\delta t$ implies convergence is achieved in one iteration and hence can be written as,

$$\frac{\partial \mu}{\partial t} = \left( \frac{\nabla \cdot \left(M \nabla \mu\right) - \sum_{\alpha=1}^{N} c^{\alpha}\left(\mu^{n}, T\right) \dfrac{\partial h_{\alpha}\left(\phi^{o}\right)}{\partial t}}{\sum_{\alpha=1}^{N} \dfrac{\partial c^{\alpha}\left(\mu^{n}, T\right)}{\partial \mu} h_{\alpha}\left(\phi^{o}\right)} \right).$$

The derived equation is identical to the binary variant of Eqn. 6.10. It is not surprising to see the similarity since we are essentially solving for the same variable $\mu$ and the difference is, while Eqn. (6.9) is an implicit type of calculation scheme of the chemical potential, the other, Eqn. (6.10) describes an explicit computation. It would be interesting to compare the performance and accuracy of both methods.

## 6.2. Asymptotic analysis

In this section we perform the asymptotic analysis of the phase-field model for a two phase binary alloy solidification with the assumption of one-sided diffusion in the liquid and vanishing diffusivity in the solid. Our aim is to derive the expressions for the kinetic coefficient in the thin interface limit for the case of solute diffusion by performing an asymptotic analysis unto second-order in the phase-field and for this purpose the analysis in

1D suffices. The asymptotic analysis is applied to the presented model ensuring no free energy excess at the interface. For simplicity, we treat here a two phase binary alloy. Hence, the chemical potential $\boldsymbol{\mu}$ will be written as $\mu$, since there exists only one independent chemical potential. At the onset, we express the grand potentials $\Psi_\alpha\left(T, \mu\right)$ as a linear expansion about the equilibrium chemical potential $\mu_{eq}$,

$$\Psi_\alpha\left(T, \mu\right) \;=\; \Psi_\alpha\left(T, \mu_{eq}\right) + \left.\frac{\partial \Psi_\alpha\left(T, \mu\right)}{\partial \mu}\right|_{\mu_{eq}} \left(\mu - \mu_{eq}\right).$$

The driving force $\Delta F^\alpha$ is then:

$$
\begin{aligned}
\Delta F^\alpha \;&=\; \left(\Psi_\alpha\left(T, \mu\right) - \Psi_\beta\left(T, \mu\right)\right) \frac{\partial h_\alpha\left(\phi\right)}{\partial \phi_\alpha} \\[2mm]
&=\; \left(\left.\frac{\partial \Psi_\alpha\left(T, \mu\right)}{\partial \mu}\right|_{\mu_{eq}} - \left.\frac{\partial \Psi_\beta\left(T, \mu\right)}{\partial \mu}\right|_{\mu_{eq}}\right) \left(\mu - \mu_{eq}\right) \frac{\partial h_\alpha\left(\phi\right)}{\partial \phi_\alpha}, \\[2mm]
&=\; -\left(c^\alpha\left(\mu_{eq}, T\right) - c^\beta\left(\mu_{eq}, T\right)\right) \left(\mu - \mu_{eq}\right) \frac{\partial h_\alpha\left(\phi\right)}{\partial \phi_\alpha},
\end{aligned}
$$

implying the evolution equation for the phase-field for a two phase system can be written as follows,

$$
\begin{aligned}
\tau_{\alpha\beta}\varepsilon^2 \frac{\partial \phi_\alpha}{\partial t} = \;&\varepsilon^2 \tilde{\sigma}_{\alpha\beta} \frac{\partial^2 \phi_\alpha}{\partial x^2} - \frac{16}{2\pi^2}\tilde{\sigma}_{\alpha\beta}\left(1 - 2\phi_\alpha\right) \\[2mm]
&+ \frac{1}{2}\varepsilon\left(c^\alpha\left(\mu_{eq}, T\right) - c^\beta\left(\mu_{eq}, T\right)\right)\left(\mu - \mu_{eq}\right)\frac{\partial h_\alpha\left(\phi_\alpha\right)}{\partial \phi_\alpha}.
\end{aligned}
\tag{6.12}
$$

Notice, we have reduced a system of two dependent equations to one independent equation, by incorporating the Lagrange multiplier formalism. Also, the interpolation function $h_\alpha\left(\phi\right)$, is now for the two phase system just a function of $\phi_\alpha$. Hence, for the forthcoming derivations, we will omit the vector notation. The interpolation functions satisfy the property, $h_\alpha\left(\phi_\alpha\right) = 1 - h_\beta\left(\phi_\beta\right)$. *Further, we consider small deviations from equilibrium which is generally a suitable assumption for most cases of solidification. For larger driving forces, such as in rapid solidification, this assumption of linearization of the driving forces will no longer hold. For the case, where* we have $D^\alpha \ll D^\beta$, the evolution equation for the chemical potential of a binary system reads,

$$\left( \frac{\partial c^\alpha (\mu, T)}{\partial \mu} h_\alpha (\phi_\alpha) + \frac{\partial c^\beta (\mu, T)}{\partial \mu} (1 - h_\alpha (\phi_\alpha)) \right) \frac{\partial \mu}{\partial t} =$$

$$\nabla \cdot \left( \left( D^\beta (1 - g_\alpha (\phi_\alpha)) \frac{\partial c^\beta (\mu, T)}{\partial \mu} \right) \nabla \mu \right) - \left( c^\alpha (\mu, T) - c^\beta (\mu, T) \right) \frac{\partial h_\alpha (\phi_\alpha)}{\partial t},$$

$$(6.13)$$

We non-dimensionalize the system of equations Eqn.(6.12) and Eqn.(6.13) by choosing the length scale $d_o = \dfrac{\tilde{\sigma}_{\alpha\beta}}{f^*}$, where $f^*$ is the energy scale of the system, the time scale $t^* = \dfrac{d_0^2}{D^\beta}$ with $D^\beta$ being the diffusivity in the liquid and replace $\tau_{\alpha\beta}$ with non-dimensionalized parameter $\zeta$ as $\dfrac{D^\beta \tau_{\alpha\beta}}{\tilde{\sigma}_{\alpha\beta}}$. The non-dimensional phase-field equation yields with the described scaling parameters,

$$\zeta \varepsilon^2 \frac{\partial \phi_\alpha}{\partial t} = \varepsilon^2 \frac{\partial^2 \phi_\alpha}{\partial x^2} - \frac{16}{2\pi^2} (1 - 2\phi_\alpha)$$

$$+ \frac{1}{2} \varepsilon \left( c^\alpha (\mu_{eq}, T) - c^\beta (\mu_{eq}, T) \right) (\mu - \mu_{eq}) \frac{\partial h_\alpha (\phi_\alpha)}{\partial \phi_\alpha},$$

while the non-dimensionalized chemical potential equation can be written as,

$$\left( \frac{\partial c^\alpha (\mu, T)}{\partial \mu} h_\alpha (\phi_\alpha) + \frac{\partial c^\beta (\mu, T)}{\partial \mu} (1 - h_\alpha (\phi_\alpha)) \right) \frac{\partial \mu}{\partial t} =$$

$$\nabla \cdot \left( (1 - g_\alpha (\phi_\alpha)) \frac{\partial c^\beta (\mu, T)}{\partial \mu} \nabla \mu \right) - \left( c^\alpha (\mu, T) - c^\beta (\mu, T) \right) \frac{\partial h_\alpha (\phi_\alpha)}{\partial t}.$$

For our further analysis, we choose the chemical potential equation for the asymptotic expansions. For the case of one-sided diffusion, it has been shown in various previous works [6, 46], that there exists a thin-interface defect called *solute trapping* when simulations are performed with interface thicknesses, orders of magnitude larger than those of a real interface. The methodology proposed to correct this effect, is the incorporation of an antitrapping current in the evolution equation of the chemical potential. While such expressions have been derived for *double well* type potentials [30, 46, 54], the case of the *double obstacle potential* is untreated so-far. We complete this gap by deriving the thin-interface limit of the model for a *double obstacle potential* and formulating an expression of the anti-trapping current $j_{at}$ for the case of one-sided diffusion. We follow the formulations described in literature and incorporate the anti-trapping term, as an additional flux of solute from the solid to the liquid in the normal direction to the interface. The modified evolution equation for the chemical potential along with the antitrapping term is,

$$\left( \frac{\partial c^{\alpha}(\mu, T)}{\partial \mu} h_{\alpha}(\phi_{\alpha}) + \frac{\partial c^{\beta}(\mu, T)}{\partial \mu} \left(1 - h_{\alpha}(\phi_{\alpha})\right) \right) \frac{\partial \mu}{\partial t} =$$
$$\nabla \cdot \left( \left(1 - g_{\alpha}(\phi_{\alpha})\right) \frac{\partial c^{\beta}(\mu, T)}{\partial \mu} \nabla \mu - j_{at} \right) - \left( c^{\alpha}(\mu, T) - c^{\beta}(\mu, T) \right) \frac{\partial h_{\alpha}(\phi_{\alpha})}{\partial t}.$$

To make sure, that the anti-trapping current appears in the first-order correction to the chemical potential we formulate the anti-trapping current of the following form,

$$j_{at} \;\; = \;\; s(\phi_{\alpha}) \, \varepsilon \left( c^{\beta}(\mu, T) - c^{\alpha}(\mu, T) \right) \frac{\partial \phi_{\alpha}}{\partial t} \frac{q_{\alpha\beta}}{|q_{\alpha\beta}|},$$

where, $s(\phi_{\alpha})$ is a function, such that the chemical potential jump vanishes at the interface. $q_{\alpha\beta}$ is the normal vector to the interface, given as $(\phi_{\alpha}\nabla\phi_{\beta} - \phi_{\beta}\nabla\phi_{\alpha})$. *To see this, use* $\phi_{\alpha} = \phi_{\beta}$, *for the case of a binary interface between the* $\alpha$ *and the* $\beta$ *interface. Then the vector* $q_{\alpha\beta}$ *reduces to* $\phi_{\alpha}\nabla(\phi_{\beta} - \phi_{\alpha})$. *Since the gradient of the scalar field* $(\phi_{\beta} - \phi_{\alpha})$ *is normal to any contour* $(\phi_{\beta} - \phi_{\alpha}) = const$, *we have* $\nabla(\phi_{\beta} - \phi_{\alpha})$ *normal to the contour* $(\phi_{\beta} - \phi_{\alpha}) = 0$ *which defines the binary interface.*
For the case of only two phases, it can be shown that the expression of the anti-trapping current can be reduced to,

$$j_{at} \;\; = \;\; -s(\phi_{\alpha}) \, \varepsilon \left( c^{\beta}(\mu, T) - c^{\alpha}(\mu, T) \right) \frac{\partial \phi_{\alpha}}{\partial t} \frac{\nabla\phi_{\alpha}}{|\nabla\phi_{\alpha}|}.$$

Note, all terms in the above equation are used in the non-dimensional form, so $\varepsilon$ is the non-dimensional parameter related to the interface width and $t$ is the non-dimensional time. Writing the phase-field and chemical potential evolution equations in one dimension, we have

$$\zeta\varepsilon^2 \frac{\partial \phi_\alpha}{\partial t} = \varepsilon^2 \frac{\partial^2 \phi_\alpha}{\partial x^2} - \frac{16}{2\pi^2}\left(1 - 2\phi_\alpha\right) +$$
$$\frac{1}{2}\varepsilon \left(c^\alpha \left(\mu_{eq}, T\right) - c^\beta \left(\mu_{eq}, T\right)\right)\left(\mu - \mu_{eq}\right)\frac{\partial h_\alpha \left(\phi_\alpha\right)}{\partial \phi_\alpha}.$$

$$\left(\frac{\partial c^\alpha \left(\mu, T\right)}{\partial \mu} h_\alpha \left(\phi_\alpha\right) + \frac{\partial c^\beta \left(\mu, T\right)}{\partial \mu}\left(1 - h_\alpha \left(\phi_\alpha\right)\right)\right)\frac{\partial \mu}{\partial t} =$$
$$\frac{\partial}{\partial x}\left(\left(1 - g_\alpha \left(\phi_\alpha\right)\right)\frac{\partial c^\beta \left(\mu, T\right)}{\partial \mu}\frac{\partial \mu}{\partial x} - s\left(\phi_\alpha\right)\varepsilon \left(c^\beta \left(\mu, T\right) - c^\alpha \left(\mu, T\right)\right)\frac{\partial \phi_\alpha}{\partial t}\right) -$$
$$\left(c^\alpha \left(\mu, T\right) - c^\beta \left(\mu, T\right)\right)\frac{\partial h_\alpha \left(\phi_\alpha\right)}{\partial t},$$

which on transformation to the moving frame (fixed to $\phi_\alpha = 0.5$) becomes,

$$-\zeta v\varepsilon^2 \frac{\partial \phi_\alpha}{\partial x} = \varepsilon^2 \frac{\partial^2 \phi_\alpha}{\partial x^2} - \frac{16}{2\pi^2}\left(1 - 2\phi_\alpha\right) +$$
$$\frac{1}{2}\varepsilon \left(c^\alpha \left(\mu_{eq}, T\right) - c^\beta \left(\mu_{eq}, T\right)\right)\left(\mu - \mu_{eq}\right)\frac{\partial h_\alpha \left(\phi_\alpha\right)}{\partial \phi_\alpha}.$$

$$-\left(\frac{\partial c^\alpha \left(\mu, T\right)}{\partial \mu} h_\alpha \left(\phi_\alpha\right) + \frac{\partial c^\beta \left(\mu, T\right)}{\partial \mu}\left(1 - h_\alpha \left(\phi_\alpha\right)\right)\right)v\frac{\partial \mu}{\partial x} =$$
$$\frac{\partial}{\partial x}\left(\left(1 - g_\alpha \left(\phi_\alpha\right)\right)\frac{\partial c^\beta \left(\mu, T\right)}{\partial \mu}\frac{\partial \mu}{\partial x} + vs\left(\phi_\alpha\right)\varepsilon \left(c^\beta \left(\mu, T\right) - c^\alpha \left(\mu, T\right)\right)\frac{\partial \phi_\alpha}{\partial x}\right) +$$
$$v\left(c^\alpha \left(\mu, T\right) - c^\beta \left(\mu, T\right)\right)\frac{\partial h_\alpha \left(\phi_\alpha\right)}{\partial x},$$

where $v$ is the non-dimensional velocity scaled as $\dfrac{V d_0}{D^\beta}$. To perform the asymptotic analysis, the region of evolution is divided into three parts. The "inner" region where there is rapid variation of the phase-field $\phi_\alpha$

and chemical potential $\mu$, and two "outer" regions which denote regions where there is little change in the phase-field $\phi_\alpha$. To probe into the inner solutions, we scale the co-ordinate with the parameter $\varepsilon$ by introducing a scaling parameter $\eta = \dfrac{x}{\varepsilon}$. With this scaling, the equations rewrite to,

$$-\zeta v \varepsilon \frac{\partial \phi_\alpha}{\partial \eta} = \frac{\partial^2 \phi_\alpha}{\partial \eta^2} - \frac{16}{2\pi^2}\left(1 - 2\phi_\alpha\right) +$$

$$\frac{1}{2}\varepsilon \left(c^\alpha\left(\mu_{eq}, T\right) - c^\beta\left(\mu_{eq}, T\right)\right)\left(\mu - \mu_{eq}\right)\frac{\partial h_\alpha\left(\phi_\alpha\right)}{\partial \phi_\alpha}.$$

$$-\left(\frac{\partial c^\alpha\left(\mu, T\right)}{\partial \mu}h_\alpha\left(\phi_\alpha\right) + \frac{\partial c^\beta\left(\mu, T\right)}{\partial \mu}\left(1 - h_\alpha\left(\phi_\alpha\right)\right)\right)\frac{v}{\varepsilon}\frac{\partial \mu}{\partial \eta} =$$

$$\frac{1}{\varepsilon^2}\frac{\partial}{\partial \eta}\left(\left(1 - g_\alpha\left(\phi_\alpha\right)\right)\frac{\partial c^\beta\left(\mu, T\right)}{\partial \mu}\frac{\partial \mu}{\partial \eta}\right) +$$

$$\frac{1}{\varepsilon}\frac{\partial}{\partial \eta}\left(vs\left(\phi_\alpha\right)\left(c^\beta\left(\mu, T\right) - c^\alpha\left(\mu, T\right)\right)\frac{\partial \phi_\alpha}{\partial \eta}\right) +$$

$$\frac{v}{\varepsilon}\left(c^\alpha\left(\mu, T\right) - c^\beta\left(\mu, T\right)\right)\frac{\partial h_\alpha\left(\phi_\alpha\right)}{\partial \eta}.$$

The strategy is to write each of the outer and inner solutions as powers of the scaling parameter $\varepsilon$ and match the outer and inner solutions order by order. The outer solutions are denoted by $\widetilde{\mu}$ and $\widetilde{\phi_\alpha}$ and are expanded by, $\widetilde{\phi_\alpha} = \widetilde{\phi_\alpha^0} + \varepsilon\widetilde{\phi_\alpha^1} + \varepsilon^2\widetilde{\phi_\alpha^2}$ and $\widetilde{\mu} = \widetilde{\mu^0} + \varepsilon\widetilde{\mu^1} + \varepsilon^2\widetilde{\mu^2}$. The inner solutions similarly writes, $\phi_\alpha = \phi_\alpha^0 + \varepsilon\phi_\alpha^1 + \varepsilon^2\phi_\alpha^2$ and $\mu = \mu^0 + \varepsilon\mu^1 + \varepsilon^2\mu^2$. The matching conditions between the outer and the inner solutions can be written by expanding each of the outer functions $\widetilde{\mu^0}, \widetilde{\mu^1}, \widetilde{\mu^2}$ as an expansion around $x = 0$, i.e $x = (0 + \eta\varepsilon)$ and equating them to the corresponding values of the inner solution: So all the derivatives are computed at the position $x = 0$, marking the interface, at $\phi_\alpha = 0.5$.

$$\lim_{\eta \to \pm\infty} \mu^0 = \widetilde{\mu^0}\big|^{\pm} \tag{6.14}$$

$$\lim_{\eta \to \pm\infty} \mu^1 = \lim_{\eta \to \pm\infty}\left(\widetilde{\mu^1}\big|^{\pm} + \eta\frac{\partial \widetilde{\mu^0}}{\partial x}\bigg|^{\pm}\right) \tag{6.15}$$

$$\lim_{\eta \to \pm\infty} \mu^2 \;=\; \lim_{\eta \to \pm\infty} \left( \widetilde{\mu^2}\big|^\pm + \eta \frac{\partial \widetilde{\mu^1}}{\partial x}\bigg|^\pm + \frac{\eta^2}{2} \frac{\partial^2 \widetilde{\mu^0}}{\partial x^2}\bigg|^\pm \right) \qquad (6.16)$$

and, the derivative matching conditions:

$$\lim_{\eta \to \pm\infty} \frac{\partial \mu^0}{\partial \eta} \;=\; 0 \qquad\qquad\qquad\qquad\qquad (6.17)$$

$$\lim_{\eta \to \pm\infty} \frac{\partial \mu^1}{\partial \eta} \;=\; \frac{\partial \widetilde{\mu^0}}{\partial x}\bigg|^\pm \qquad\qquad\qquad (6.18)$$

$$\lim_{\eta \to \pm\infty} \frac{\partial \mu^2}{\partial \eta} \;=\; \lim_{\eta \to \pm\infty} \left( \frac{\partial \widetilde{\mu^1}}{\partial x}\bigg|^\pm + \eta \frac{\partial^2 \widetilde{\mu^0}}{\partial x^2}\bigg|^\pm \right). \qquad (6.19)$$

The matching conditions for the phase-field are trivial, as the phase-field is constant in the bulk on both sides. Hence the outer solution in the phase-field is non-zero only for the lowest order. Now we solve the phase-field and chemical potential equations order by order and derive the various boundary conditions for the chemical potential as solvability conditions.

## 6.2.1. Sharp interface limit

The phase-field equation at zero order in $\varepsilon$ reads,

$$\frac{\partial^2 \phi_\alpha^0}{\partial \eta^2} - \frac{16}{2\pi^2}\left(1 - 2\phi_\alpha^0\right) = 0.$$

Integrating yields, $\dfrac{\partial \phi_\alpha^0}{\partial \eta} = -\dfrac{4}{\pi}\sqrt{\phi_\alpha^0\left(1 - \phi_\alpha^0\right)}$, where the sign results from the boundary conditions, $\lim_{\eta \to +\infty} \phi_\alpha^0 = 0$ and $\lim_{\eta \to -\infty} \phi_\alpha^0 = 1$. The lowest order chemical potential equation at order $1/\varepsilon^2$ is,

$$\frac{1}{\varepsilon^2} \frac{\partial}{\partial \eta} \left( \left(1 - g_\alpha\left(\phi_\alpha^0\right)\right) \frac{\partial c^\beta\left(\mu^0, T\right)}{\partial \mu} \frac{\partial \mu^0}{\partial \eta} \right) = 0.$$

Integrating, the above equation once we get,

$$\left(1 - g_\alpha\left(\phi_\alpha^0\right)\right) \frac{\partial c^\beta\left(\mu^0, T\right)}{\partial \mu} \frac{\partial \mu^0}{\partial \eta} \;=\; A_1. \tag{6.20}$$

We observe $\lim_{\eta \to \infty} g_\alpha\left(\phi_\alpha^0\right) = 0$ and the factor $\dfrac{\partial c^\beta\left(\mu^0, T\right)}{\partial \mu}$ is non-zero. Using the matching condition in Eqn. (6.17), we derive that $A_1$ is zero. Inserting $A_1 = 0$ into Eqn. (6.20) and integrating once we get the following,

$$\mu^0 \;=\; \overline{\mu^0}.$$

$\overline{\mu^0}$ is an integration constant. To fix the value, we insert this constant in the phase-field equation at order $\varepsilon$ that reads,

$$-\zeta v \frac{\partial \phi_\alpha^0}{\partial \eta} = \frac{\partial^2 \phi_\alpha^1}{\partial \eta^2} + \frac{16}{\pi^2}\phi_\alpha^1 + \frac{1}{2}\left(c^\alpha\left(\mu_{eq}, T\right) - c^\beta\left(\mu_{eq}, T\right)\right)\left(\mu^0 - \mu_{eq}\right)\frac{\partial h_\alpha\left(\phi_\alpha^0\right)}{\partial \phi_\alpha}.$$

For brevity we determine the constant $\left(\mu^0 - \mu_{eq}\right)$ from the solvability condition. It is useful to identify an useful operator which derives from the phase-field equation at zeroth order as follows,

$$\frac{\partial^2 \phi_\alpha^0}{\partial \eta^2} - \frac{16}{2\pi^2}\left(1 - 2\phi_\alpha^0\right) \;=\; 0.$$

Differentiating, the above equation and re-arranging we get,

$$\left(\frac{\partial^2}{\partial \eta^2} + \frac{16}{\pi^2}\right)\frac{\partial \phi_\alpha^0}{\partial \eta} = 0. \tag{6.21}$$

The term in the brackets $\dfrac{\partial^2}{\partial \eta^2} + \dfrac{16}{\pi^2}$ is a linear operator and we define this as $L$. Using this linear operator, the phase-field equation at order $\varepsilon$ is,

$$L\phi_\alpha^1 = -\zeta v \frac{\partial \phi_\alpha^0}{\partial \eta} - \frac{1}{2}\left(c^\alpha\left(\mu_{eq}, T\right) - c^\beta\left(\mu_{eq}, T\right)\right)\left(\mu^0 - \mu_{eq}\right)\frac{\partial h_\alpha\left(\phi_\alpha^0\right)}{\partial \phi_\alpha}.$$

From Eqn. (6.21), we see that $\dfrac{\partial \phi_\alpha^0}{\partial \eta}$ is a homogeneous solution of the operator $L$, hence the solvability condition for a non-trivial $\phi_\alpha^1$ reads,

$$\int_{-\infty}^{\infty} -\zeta v \left(\frac{\partial \phi_\alpha^0}{\partial \eta}\right)^2 \partial \eta =$$

$$\int_{-\infty}^{\infty} \frac{1}{2} \left(c^\alpha\left(\mu_{eq}, T\right) - c^\beta\left(\mu_{eq}, T\right)\right)\left(\mu^0 - \mu_{eq}\right) \frac{\partial h_\alpha\left(\phi_\alpha^0\right)}{\partial \phi_\alpha} \frac{\partial \phi_\alpha^0}{\partial \eta} \partial \eta.$$

Making use of the integrals $\int_{-\infty}^{\infty} \left(\frac{\partial \phi_\alpha^0}{\partial \eta}\right)^2 \partial \eta = \frac{1}{2}$, and $\int_{-\infty}^{\infty} \frac{\partial h_\alpha\left(\phi_\alpha^0\right)}{\partial \phi_\alpha} \partial \eta = -1$, the equation simplifies to

$$\left(\mu^0 - \mu_{eq}\right) \;=\; \frac{-\zeta v}{\left(c^\beta\left(\mu_{eq}, T\right) - c^\alpha\left(\mu_{eq}, T\right)\right)} \tag{6.22}$$

This is the departure from the equilibrium chemical potential in the *sharp interface limit*.

## 6.2.2. Thin interface limit

For, the thin interface correction we solve the chemical potential equation at the next order at $1/\varepsilon$,

$$-\left(\frac{\partial c^\alpha\left(\mu^0, T\right)}{\partial \mu} h_\alpha\left(\phi_\alpha^0\right) + \frac{\partial c^\beta\left(\mu^0, T\right)}{\partial \mu}\left(1 - h_\alpha\left(\phi_\alpha^0\right)\right)\right) \frac{v}{\varepsilon} \frac{\partial \mu^0}{\partial \eta} =$$

$$\frac{1}{\varepsilon} \frac{\partial}{\partial \eta}\left(\left(1 - g_\alpha\left(\phi_\alpha^0\right)\right) \frac{\partial c^\beta\left(\mu^0, T\right)}{\partial \mu} \frac{\partial \mu^1}{\partial \eta}\right) +$$

$$\frac{1}{\varepsilon} \frac{\partial}{\partial \eta}\left(v s\left(\phi_\alpha^0\right)\left(c^\beta\left(\mu^0, T\right) - c^\alpha\left(\mu^0, T\right)\right) \frac{\partial \phi_\alpha^0}{\partial \eta}\right) +$$

$$\frac{v}{\varepsilon}\left(c^\alpha\left(\mu^0, T\right) - c^\beta\left(\mu^0, T\right)\right) \frac{\partial h_\alpha\left(\phi_\alpha^0\right)}{\partial \eta}. \tag{6.23}$$

Note, at order $1/\varepsilon$, there are additional terms. However we make use of the fact that $\mu^0$ is constant and hence all its derivatives vanish, so that Eqn. (6.23) simplifies to,

$$
\frac{\partial}{\partial \eta}\left(\left(1 - g_\alpha\left(\phi_\alpha^0\right)\right) \frac{\partial c^\beta\left(\mu^0, T\right)}{\partial \mu} \frac{\partial \mu^1}{\partial \eta}\right) =
$$

$$
- v\frac{\partial}{\partial \eta}\left(s\left(\phi_\alpha^0\right)\left(c^\beta\left(\mu^0, T\right) - c^\alpha\left(\mu^0, T\right)\right) \frac{\partial \phi_\alpha^0}{\partial \eta}\right) +
$$

$$
v\left(c^\beta\left(\mu^0, T\right) - c^\alpha\left(\mu^0, T\right)\right) \frac{\partial h_\alpha\left(\phi_\alpha^0\right)}{\partial \eta}.
$$

Integrating this once we get,

$$
\left(\left(1 - g_\alpha\left(\phi_\alpha^0\right)\right) \frac{\partial c^\beta\left(\mu^0, T\right)}{\partial \mu} \frac{\partial \mu^1}{\partial \eta}\right) =
$$

$$
- v\left(s\left(\phi_\alpha^0\right)\left(c^\beta\left(\mu^0, T\right) - c^\alpha\left(\mu^0, T\right)\right) \frac{\partial \phi_\alpha^0}{\partial \eta}\right) +
$$

$$
v\left(c^\beta\left(\mu^0, T\right) - c^\alpha\left(\mu^0, T\right)\right) h_\alpha\left(\phi_\alpha^0\right) + A_2.
$$

To fix $A_2$ we take $\lim_{\eta->-\infty}$ which gives, $\left(1 - g_\alpha\left(\phi_\alpha\right)\right) \to 0$, $\dfrac{\partial c^\beta\left(\mu^0, T\right)}{\partial \mu}$ is a positive constant, $\dfrac{\partial \phi_\alpha^0}{\partial \eta} \to 0$ and $h_\alpha\left(\phi_\alpha^0\right) \to 1$. Therefore, the value of $A_2 =$ $- v\left(c^\beta\left(\mu^0, T\right) - c^\alpha\left(\mu^0, T\right)\right)$. Substituting this in the above equation and re-arranging we get,

$$
\frac{\partial \mu^1}{\partial \eta} = \frac{v\left(c^\beta\left(\mu^0, T\right) - c^\alpha\left(\mu^0, T\right)\right)\left(h_\alpha\left(\phi_\alpha^0\right) - 1 - s\left(\phi_\alpha^0\right) \dfrac{\partial \phi_\alpha^0}{\partial \eta}\right)}{\dfrac{\partial c^\beta\left(\mu^0, T\right)}{\partial \mu}\left(1 - g_\alpha\left(\phi_\alpha^0\right)\right)}.
$$

For brevity we denote the expression $\dfrac{h_\alpha\left(\phi_\alpha^0\right) - 1 - s\left(\phi_\alpha^0\right)\dfrac{\partial \phi_\alpha^0}{\partial \eta}}{\left(1 - g_\alpha\left(\phi_\alpha^0\right)\right)}$ as $r\left(\phi_\alpha^0\right)$.
Substituting this in the preceding equation and integrating we get,

$$\mu^1 = \overline{\mu^1} + \frac{v\left(c^\beta\left(\mu^0, T\right) - c^\alpha\left(\mu^0, T\right)\right)}{\dfrac{\partial c^\beta\left(\mu^0, T\right)}{\partial \mu}}\int_0^\eta r\left(\phi_\alpha^0\right)\partial \eta. \qquad (6.24)$$

To obtain the integration constant $\overline{\mu^1}$, we write the phase-field equation at order $\varepsilon^2$,

$$-\zeta v \varepsilon^2 \frac{\partial \phi_\alpha^1}{\partial \eta} = \varepsilon^2 \frac{\partial^2 \phi_\alpha^2}{\partial \eta^2} +$$

$$\varepsilon^2 \frac{16}{\pi^2}\phi_\alpha^2 + \frac{1}{2}\varepsilon^2 \phi_\alpha^1 \left(c^\alpha\left(\mu_{eq}, T\right) - c^\beta\left(\mu_{eq}, T\right)\right)\left(\mu^0 - \mu_{eq}\right)\frac{\partial^2 h_\alpha\left(\phi_\alpha^0\right)}{\partial \phi_\alpha^2} +$$

$$\frac{1}{2}\varepsilon^2 \left(c^\alpha\left(\mu_{eq}, T\right) - c^\beta\left(\mu_{eq}, T\right)\right)\mu^1 \frac{\partial h_\alpha\left(\phi_\alpha^0\right)}{\partial \phi_\alpha} \quad \text{or,}$$

$$L\phi_\alpha^2 = -\zeta v \frac{\partial \phi_\alpha^1}{\partial \eta}$$

$$-\frac{1}{2}\left(c^\alpha\left(\mu_{eq}, T\right) - c^\beta\left(\mu_{eq}, T\right)\right)\left[\phi_\alpha^1\left(\mu^0 - \mu_{eq}\right)\frac{\partial^2 h_\alpha\left(\phi_\alpha^0\right)}{\partial \phi_\alpha^2} + \mu^1 \frac{\partial h_\alpha\left(\phi_\alpha^0\right)}{\partial \phi_\alpha}\right].$$

The solvability condition for a non-trivial $\phi_\alpha^2$ can be derived as,

$$-\int_{-\infty}^{\infty} \zeta v \frac{\partial \phi_\alpha^1}{\partial \eta}\frac{\partial \phi_\alpha^0}{\partial \eta}\partial \eta -$$

$$\int_{-\infty}^{\infty} \frac{1}{2}\phi_\alpha^1\left(c^\alpha\left(\mu_{eq}, T\right) - c^\beta\left(\mu_{eq}, T\right)\right)\left(\mu^0 - \mu_{eq}\right)\frac{\partial^2 h_\alpha\left(\phi_\alpha^0\right)}{\partial \phi_\alpha^2}\frac{\partial \phi_\alpha^0}{\partial \eta}\partial \eta -$$

$$\int_{-\infty}^{\infty} \frac{1}{2}\left(c^\alpha\left(\mu_{eq}, T\right) - c^\beta\left(\mu_{eq}, T\right)\right)\mu^1 \frac{\partial h_\alpha\left(\phi_\alpha^0\right)}{\partial \phi_\alpha}\frac{\partial \phi_\alpha^0}{\partial \eta}\partial \eta = 0. \qquad (6.25)$$

To see the nature of the first integral, we make use of the fact that $\phi_\alpha^1$ satisfies the phase-field equation at order $\varepsilon$ which reads,

$$L\phi_\alpha^1 = -\zeta v \frac{\partial \phi_\alpha^0}{\partial \eta} - \frac{1}{2}\left(c^\alpha\left(\mu_{eq}, T\right) - c^\beta\left(\mu_{eq}, T\right)\right)\left(\mu^0 - \mu_{eq}\right)\frac{\partial h_\alpha\left(\phi_\alpha^0\right)}{\partial \phi_\alpha} \quad (6.26)$$

The phase-field profile $\phi_\alpha^0$ is part of a *sinus* curve and hence is an odd-function, which implies its derivative $\dfrac{\partial \phi_\alpha^0}{\partial \eta}$ is even. Similarly, the interpolation function $h_\alpha\left(\phi_\alpha^0\right)$ is an odd function and hence the function $\dfrac{\partial h_\alpha\left(\phi_\alpha^0\right)}{\partial \phi_\alpha}$ is an *even* function. In order to realize this, we utilize the anti-symmetric property of the interpolation function with respect to the $\eta = 0$, and equivalently where $\phi_\alpha = 1/2$,

$$h_\alpha\left(\phi_\alpha^0\left(\eta\right)\right) - \frac{1}{2} = \frac{1}{2} - h_\alpha\left(\phi_\alpha^0\left(-\eta\right)\right).$$

Differentiating both sides with respect to $\eta$ and using the *even* property of $\dfrac{\partial \phi_\alpha^0}{\partial \eta}$ we derive,

$$\frac{\partial h_\alpha\left(\phi_\alpha\left(\eta\right)\right)}{\partial \phi_\alpha} = \frac{\partial h_\alpha\left(\phi_\alpha\left(-\eta\right)\right)}{\partial \phi_\alpha},$$

which proves $\dfrac{\partial h_\alpha\left(\phi_\alpha\left(\eta\right)\right)}{\partial \phi_\alpha}$ is even. Conversely, differentiating again implies the second derivative $\dfrac{\partial^2 h_\alpha\left(\phi_\alpha\left(\eta\right)\right)}{\partial \phi_\alpha^0{}^2}$ is odd. Using these properties we directly find that the R.H.S of equation Eqn. (6.26) is even. Combined with the fact that the operator $L$ is of the form $\dfrac{\partial^2}{\partial \eta^2} + \dfrac{16}{\pi^2}$, which does not modify the characteristics of the R.H.S, we derive that $\phi_\alpha^1$ is even. Putting all the arguments together results in the implication that only the second integral with the term $\mu^1$ in the solvability condition Eqn. (6.25) does not vanish and the solvability condition simplifies to,

$$\int_{-\infty}^{\infty} \frac{1}{2}\left(c^\alpha\left(\mu_{eq}, T\right) - c^\beta\left(\mu_{eq}, T\right)\right)\mu^1 \frac{\partial h_\alpha\left(\phi_\alpha^0\right)}{\partial \phi_\alpha}\frac{\partial \phi_\alpha^0}{\partial \eta}\partial \eta = 0.$$

Inserting Eqn.(6.24) for $\mu^1$ into the above solvability condition, we derive an equation for $\overline{\mu^1}$ given by,

$$\overline{\mu^1} = \frac{v\left(c^\beta\left(\mu^0,T\right) - c^\alpha\left(\mu^0,T\right)\right)}{\dfrac{\partial c^\beta\left(\mu^0,T\right)}{\partial\mu}} \underbrace{\int_{-\infty}^{\infty}\left[\int_0^{\eta} r\left(\phi_\alpha^0\right)\partial\eta\right]\frac{\partial h_\alpha\left(\phi_\alpha^0\right)}{\partial\phi_\alpha}\frac{\partial\phi_\alpha^0}{\partial\eta}\partial\eta}_{:=\tilde{M}}\,.$$

With the short hand notation $\tilde{M}$ we can write,

$$\mu^1 = \frac{\left(c^\beta\left(\mu^0,T\right) - c^\alpha\left(\mu^0,T\right)\right)}{\dfrac{\partial c^\beta\left(\mu^0,T\right)}{\partial\mu}} v\left(\tilde{M} + \int_0^{\eta} r\left(\phi_\alpha^0\right)\partial\eta\right)\,.$$

The thin-interface limit which denotes the macroscopic chemical potential at first order $\widetilde{\mu_1}$ can be derived by using the $\lim_{\eta\to\pm\infty}$ and the matching condition in Eqn. (6.18) and giving,

$$\begin{aligned}
\lim_{\eta\to\pm\infty}\frac{\partial\mu^1}{\partial\eta} &= \left.\frac{\partial\widetilde{\mu^0}}{\partial x}\right|^{\pm}\\
&= \frac{\left(c^\beta\left(\mu^0,T\right) - c^\alpha\left(\mu^0,T\right)\right)}{\dfrac{\partial c^\beta\left(\mu^0,T\right)}{\partial\mu}} vr\left(\phi_\alpha^0\pm\right),
\end{aligned}$$

where $\phi_\alpha^0\pm$ denotes the value of $\phi_\alpha^0$ at the respective bulk sides. Employing the matching condition given by Eqn.(6.15) we have,

$$\begin{aligned}
\left.\widetilde{\mu^1}\right|^{\pm} &= \lim_{\eta\to\pm\infty}\left(\mu^1 - \eta\left.\frac{\partial\widetilde{\mu^0}}{\partial x}\right|^{\pm}\right)\\
&= \frac{\left(c^\beta\left(\mu^0,T\right) - c^\alpha\left(\mu^0,T\right)\right)}{\dfrac{\partial c^\beta\left(\mu^0,T\right)}{\partial\mu}} v\left(\tilde{M} + \tilde{F}^{\pm}\right),
\end{aligned} \qquad (6.27)$$

where we define $\left.\tilde{F}\right|^{\pm}$ as follows,

$$\tilde{F}^{\pm} = \int_0^{\pm\infty}\left(r\left(\phi_\alpha^0\right) - r\left(\phi_\alpha^0\pm\right)\right)\partial\eta.$$

We realize that the limits on both sides(solid and liquid) do not match if $\tilde{F}^+ \neq \tilde{F}^-$ which gives rise to a *chemical potential jump* at the interface. To remove this jump one must now, devise a way to make the following condition true,

$$\int_0^\infty \left(r\left(\phi_\alpha^0\right) + 1\right) \partial\eta = \int_0^{-\infty} r\left(\phi_\alpha^0\right) \partial\eta,$$

where we have made use of the fact that $\dfrac{\partial\phi_\alpha^0}{\partial\eta}$ is zero at $\eta \to \pm\infty$ and $h_\alpha\left(\phi_\alpha^0\right) - 1 \to 0$ at $\eta = -\infty$ and $h\left(\phi_\alpha^0\right) = 0$ at $\eta = +\infty$. We notice, that these are properties directly related to our interpolation function $h_\alpha\left(\phi_\alpha^0\right)$. We intend to retrieve the same properties (implying $r\left(\phi_\alpha\right) = h\left(\phi_\alpha\right) - 1$), which is a reasonable choice, then we get $s\left(\phi_\alpha^0\right)$ as,

$$s\left(\phi_\alpha^0\right) = -\frac{g_\alpha\left(\phi_\alpha^0\right)\left(1 - h\left(\phi_\alpha^0\right)\right)}{\dfrac{\partial\phi_\alpha^0}{\partial\eta}}. \tag{6.28}$$

With this modification we define $\tilde{F} := \tilde{F}^+ \left(= \tilde{F}^-\right)$ and the macroscopic chemical potential $\widetilde{\mu_1}$ at first order in Eqn. (6.27) yields,

$$\begin{aligned}
\widetilde{\mu^1}\big|^\pm &= \lim_{\eta\to\pm\infty}\left(\mu^1 - \eta\frac{\partial\widetilde{\mu^0}}{\partial x}\bigg|^\pm\right) \\
&= \frac{\left(c^\beta\left(\mu^0, T\right) - c^\alpha\left(\mu^0, T\right)\right)}{\dfrac{\partial c^\beta\left(\mu^0, T\right)}{\partial\mu}} v\left(\tilde{M} + \tilde{F}\right).
\end{aligned}$$

and the chemical potential until the first order in $\varepsilon$ writes $\widetilde{\mu}\big|^\pm = \mu^0 + \varepsilon\widetilde{\mu^1}\big|^\pm$, which upon subtracting $\mu_{eq}$ from both sides becomes,

$$\widetilde{\mu}\big|^\pm - \mu_{eq} = \left(\mu^0 - \mu_{eq}\right) + \varepsilon\frac{\left(c^\beta\left(\mu^0, T\right) - c^\alpha\left(\mu^0, T\right)\right)}{\dfrac{\partial c^\beta\left(\mu^0, T\right)}{\partial\mu}} v\left(\tilde{M} + \tilde{F}\right).$$

Putting all physical properties in their respective dimensions, we get,

$$\tilde{\mu}|^{\pm} - \mu_{eq} = \left(\mu^0 - \mu_{eq}\right) + \varepsilon \frac{\left(c^{\beta}\left(\mu^0, T\right) - c^{\alpha}\left(\mu^0, T\right)\right)}{\left(D^{\beta}\right)\dfrac{\partial c^{\beta}\left(\mu^0, T\right)}{\partial \mu}} V\left(\tilde{M} + \tilde{F}\right), \quad (6.29)$$

where $\varepsilon$ is hereafter, in dimensions of length.

## 6.2.3. Kinetic coefficient and the antitrapping current

To relate the total departure from equilibrium at first order in $\varepsilon$ given in Eqn. (6.29) it is customary to write the modified temperature of the interface $T$ due to the Gibbs-Thomson effect written as,

$$T = T_m - |m_\beta| c_i^{\beta} - \Gamma \kappa - \overline{\beta} V, \qquad (6.30)$$

where $T$ and $c_i^{\beta}$ are the interfacial temperatures and the concentrations of the liquid, while $m_\beta$ is the slope of the liquidus and $T_m$ is the melting point of the pure component. $\overline{\beta}$ is the kinetic coefficient and $\Gamma$ is the Gibbs-Thomson coefficient. Then $T$ can be written as follows:

$$T = T_m - |m_\beta| c^{\beta}\left(\mu_{eq}, T\right). \qquad (6.31)$$

With this, the Gibbs-Thompson equation is modified as,

$$\left(T_m - |m_\beta| c_i^{\beta}\right) - \left(T_m - |m_\beta| c^{\beta}\left(\mu_{eq}, T\right)\right) = \Gamma \kappa + \overline{\beta} V \qquad (6.32)$$

The first bracketed term on the left hand side is the modified melting point of the interface due to constitutional undercooling because of the shift of interfacial concentration $c_i^{\beta}$ with respect to the equilibrium liquidus concentration at this temperature $c^{\beta}\left(\mu_{eq}, T\right)$. The second bracketed term is the temperature of solidification. Their difference is nothing but the equivalent undercooling $\Delta T$, which matches the Gibbs-Thomson equation of a pure material. Since we are here treating only one dimensional problems, curvature undercooling drops out and the effective undercooling reads,

$$\Delta T = m_\beta \left(c^{\beta}\left(\mu_{eq}, T\right) - c_i\right) = \overline{\beta} V. \qquad (6.33)$$

In order to relate the undercooling at the interface, $\widetilde{\Delta T}$ which is the macroscopic undercooling at first order, to the deviation of the macroscopic chemical potential from equilibrium in the thin-interface limit, we multiply Eqn. (6.29) by $m_\beta \dfrac{\partial c^\beta\left(\mu_{eq}, T\right)}{\partial \mu}$ such that the left hand side of the equation is nothing but $m_\beta \left(c_i^\beta - c^\beta\left(\mu_{eq}, T\right)\right)$ and alternatively $-\widetilde{\Delta T}$. With this modification, the total Eqn. (6.29) becomes,

$$-\widetilde{\Delta T}\big|^\pm = m_\beta \frac{\partial c^\beta\left(\mu_{eq}, T\right)}{\partial \mu}\left[\left(\mu^0 - \mu_{eq}\right) + \right.$$
$$\left. \varepsilon\frac{\left(c^\beta\left(\mu^0, T\right) - c^\alpha\left(\mu^0, T\right)\right)}{\left(D^\beta\right)\dfrac{\partial c^\beta\left(\mu^0, T\right)}{\partial \mu}} V\left(\tilde{M} + \tilde{F}\right)\right]. \qquad (6.34)$$

Using the result obtained in Eqn. (6.22), inserting the appropriate dimensions and substituting the relation between the undercooling and the kinetic coefficient, $\left(\widetilde{\Delta T}\big|^\pm = \widetilde{\beta}\big|^\pm V\right)$ we derive the equation of the kinetic coefficient $\left(\widetilde{\beta}\big|^\pm = \widetilde{\beta}\right)$ as,

$$\widetilde{\beta} = \frac{m_\beta \dfrac{\partial c^\beta\left(\mu_{eq}, T\right)}{\partial \mu}}{\left(c^\beta\left(\mu_{eq}, T\right) - c^\alpha\left(\mu_{eq}, T\right)\right)} \times$$
$$\left[\tau_{\alpha\beta} - \varepsilon\frac{\left(c^\beta\left(\mu^0, T\right) - c^\alpha\left(\mu^0, T\right)\right)\left(c^\beta\left(\mu_{eq}, T\right) - c^\alpha\left(\mu_{eq}, T\right)\right)}{\left(D^\beta\right)\dfrac{\partial c^\beta\left(\mu^0, T\right)}{\partial \mu}}\left(\tilde{M} + \tilde{F}\right)\right].$$
$$(6.35)$$

Now we make the approximation $\left(c^\beta\left(\mu^0, T\right) - c^\alpha\left(\mu^0, T\right)\right) \approx \left(c^\beta\left(\mu_{eq}, T\right) - c^\alpha\left(\mu_{eq}, T\right)\right)$, which is valid for small driving forces. Utilizing the approximation, the expression for the kinetic coefficient until the first order becomes,

$$\widetilde{\overline{\beta}} = \frac{m_\beta \dfrac{\partial c^\beta\left(\mu_{eq}, T\right)}{\partial \mu}}{\left(c^\beta\left(\mu_{eq}, T\right) - c^\alpha\left(\mu_{eq}, T\right)\right)} \left[\tau_{\alpha\beta} - \varepsilon \frac{\left(c^\beta\left(\mu_{eq}, T\right) - c^\alpha\left(\mu_{eq}, T\right)\right)^2}{\left(D^\beta\right)\dfrac{\partial c^\beta\left(\mu^0, T\right)}{\partial \mu}} \left(\tilde{M} + \tilde{F}\right)\right]$$

$$(6.36)$$

An alternative form can also be written using some basic thermodynamics relating $\dfrac{\partial c^\beta\left(\mu_{eq}, T\right)}{\partial \mu}$ with the latent heat of transformation $L_\alpha$ using the Clausius-Clapeyron equation for alloys which writes $\dfrac{d\mu_{eq}}{\partial T} = \dfrac{\partial \mu_{eq}}{\partial c^\beta}\dfrac{dc^\beta\left(\mu_{eq}, T\right)}{dT} = \dfrac{L^\alpha}{T\left(c^\beta\left(\mu_{eq}, T\right) - c^\alpha\left(\mu_{eq}, T\right)\right)}$. Using $m_\beta = \dfrac{dc^\beta\left(\mu_{eq}, T\right)}{dT}$, we derive, $\dfrac{m_\beta \dfrac{\partial c^\beta\left(\mu_{eq}, T\right)}{\partial \mu}}{\left(c^\beta\left(\mu_{eq}, T\right) - c^\alpha\left(\mu_{eq}, T\right)\right)} = \dfrac{T}{L^\alpha}$. This gives an equivalent form for the kinetic coefficient as,

$$\widetilde{\overline{\beta}} = \frac{T}{L^\alpha} \left[\tau_{\alpha\beta} - \varepsilon \frac{\left(c^\beta\left(\mu_{eq}, T\right) - c^\alpha\left(\mu_{eq}, T\right)\right)^2}{\left(D^\beta\right)\dfrac{\partial c^\beta\left(\mu^0, T\right)}{\partial \mu}} \left(\tilde{M} + \tilde{F}\right)\right].$$

$$(6.37)$$

From this it is easy to see, that to perform simulations with vanishing interface kinetic coefficient $\left(\widetilde{\overline{\beta}} = 0\right)$, one can choose the relaxation constant $\tau_{\alpha\beta}$ according to the relation,

$$\tau_{\alpha\beta} = \varepsilon \frac{\left(c^\beta\left(\mu_{eq}, T\right) - c^\alpha\left(\mu_{eq}, T\right)\right)^2}{\left(D^\beta\right)\dfrac{\partial c^\beta\left(\mu^0, T\right)}{\partial \mu}} \left(\tilde{M} + \tilde{F}\right).$$

$$(6.38)$$

For the typical interpolation polynomials of cubic and quartic type polynomial, when used in combination with the obstacle potential, the values of $\tilde{F}$ and $\tilde{M}$ are tabulated below,

| | $\tilde{M}$ | $\tilde{F}$ |
|---|---|---|
| $h\left(\phi_\alpha\right) = \phi_\alpha^2\left(3 - 2\phi_\alpha\right)$ | 0.063828 | 0.158741 |
| $h\left(\phi_\alpha\right) = \phi_\alpha^3\left(10 - 15\phi_\alpha + 6\phi_\alpha^2\right)$ | 0.052935 | 0.129288 |

Finally, the anti-trapping current along with the derived $s\left(\phi_\alpha^0\right)$ in Eqn. (6.28) is given by,

$$
j_{at} = -\frac{\pi\varepsilon}{4}\frac{g_\alpha\left(\phi_\alpha^0\right)\left(1 - h_\alpha\left(\phi_\alpha^0\right)\right)}{\sqrt{\phi_\alpha^0\left(1 - \phi_\alpha^0\right)}}\left(c^\beta\left(\mu^0, T\right) - c^\alpha\left(\mu^0, T\right)\right)\frac{\partial\phi_\alpha}{\partial t}\frac{\nabla\phi_\alpha}{|\nabla\phi_\alpha|}.
$$

$$(6.39)$$

It in interesting to see the similarity between the equations Eqn. (6.38) and Eqn. (5.29), used for deriving vanishing interface kinetics for the case of solute diffusion and pure thermal diffusion respectively. The similarity can be appreciated if we compare the evolution equations for the temperature field and the chemical potential. The rejection of solute $c^\beta\left(\mu_{eq}, T\right) - c^\alpha\left(\mu_{eq}, T\right)$ is analogous to the rejection of latent heat $L_\alpha$. Correspondingly, the mobility in the case of the thermal diffusion is $K$, while it is $D^\beta\dfrac{\partial c^\beta}{\partial\mu}$ for the case of solute diffusion. The resemblance is not surprising since, both are diffusion equations, involving phase change at the interface.

## 6.2.4. Effect of curvature and anisotropy

With the above analysis, we derive the expressions for the relaxation constant and the antitrapping current, which are dependent on the chemical potential at the zeroth order, $\mu^0$. While in one dimensional problems, its value depends on the local normal velocity and can be determined by the Eqn. (6.22), in the presence of curvature, the Gibbs-Thomson condition is modified through the contribution of the term proportional to $\sigma\kappa$ which modifies the sharp interface limit for isotropic surface energies, given in Eqn.(6.22) in dimensional units as,

$$
\left(\mu^0 - \mu_{eq}\right) = \frac{\tau_{\alpha\beta}V}{\left(c^\alpha\left(\mu_{eq}, T\right) - c^\beta\left(\mu_{eq}, T\right)\right)} + \frac{\sigma\kappa}{c^\alpha\left(\mu_{eq}, T\right) - c^\beta\left(\mu_{eq}, T\right)}
$$

The preceding equation can be derived by considering the extra term arising from writing the laplacian in curvilinear co-ordinates represented using the curvature $\kappa$ and the arc length as in [30, 50]. The value of the chemical potential $\mu^0$ derived through the preceding expression, is difficult to utilize in the expressions derived for the kinetic coefficient and the anti-trapping current, since the values of the curvature and the velocity are not known a priori. A work around this problem would be to use the approximation, $\dfrac{\partial c^\beta\left(\mu^0, T\right)}{\partial \mu} \approx \dfrac{\partial c^\beta\left(\mu_{eq}, T\right)}{\partial \mu}$ which is valid for small departures from equilibrium relevant for most phase transition problems occurring at lower undercoolings in the absence of appreciable interface kinetics. The same approximation can also be applied for the the rejection, $c^\alpha\left(\mu^0, T\right) - c^\beta\left(\mu^0, T\right)$ appearing in both the expressions for the relaxation constant and the anti-trapping current, which varies little from its value at at the equilibrium chemical potential for lower undercoolings.

For larger departures from equilibrium we propose to dynamically evaluate the expressions for the relaxation constant $\tau_{\alpha\beta}$ and the anti-trapping current. To do this we need the chemical potential in the sharp interface limit which is the average value across the interface. Since this is computationally time consuming to evaluate, we use the local chemical potential for the dynamic computation of the above mentioned quantities. This introduces an error of order $O(\varepsilon^2)$ and higher for the anti-trapping which can be realized by expanding the term $c^\beta\left(\mu^0, T\right) - c^\alpha\left(\mu^0, T\right)$ around the local chemical potential $\mu$. The highest order correction would be proportional to $\left(\dfrac{\partial c^\beta}{\partial \mu} - \dfrac{\partial c^\alpha}{\partial \mu}\right)\left(\mu^0 - \mu\right)$, where $\mu^0 - \mu$ is at highest order proportional to $O(\varepsilon)$ rendering the leading order correction due to this implementation of the antitrapping current proportional to $O(\varepsilon^2)$. A similar result can be derived for the case of the relaxation constant $\tau_{\alpha\beta}$. Since, in the thin-interface limit, we only claim to derive the relations with accuracy of order $O(\varepsilon)$, this scheme should be certainly acceptable. Another point worth mentioning regards the treatment of anisotropy in kinetics. While using Eqn.(6.38), problems with vanishing interface kinetics in isotropic situations can be treated, in order to achieve interface evolution with non-vanishing interface kinetics in the case of isotropic system, would require the back calculation of the relaxation constant through the Eqn.(6.37). The more realistic situation of anisotropy in surface energy and kinetics can be treated through a modification of Eqn.(6.38). We follow a route suggested in [48, 50] of a simple derivation, by writing the equations of motion for the normal direction but excluding curvature. The anisotropy in the surface energy is affected by writing the gradient energy contribution as $\gamma_0 a_c \left(\mathbf{n}\right)^2 \left|q_{\alpha\beta}\right|^2$, where $\mathbf{n}$ is the unit normal vector to the interface defined as $\dfrac{q_{\alpha\beta}}{\left|q_{\alpha\beta}\right|}$ and $a_c\left(\mathbf{n}\right)$ describes the anisotropy in the surface energy. A similar function $\tau_{\alpha\beta}\left(\mathbf{n}\right)$ is used for tailoring the anisotropy in the kinetic coefficient. The major modification in the asymptotics through this calculation, is the transformation of the gradient in the phase-field profile at leading order which becomes, $\dfrac{\partial \phi_\alpha^0}{\partial \eta} = -\dfrac{1}{a_c\left(\mathbf{n}\right)} \dfrac{4}{\pi} \sqrt{\phi_\alpha^0 \left(1 - \phi_\alpha^0\right)}$. Incorporating this result in the asymptotics, yields the following expression for the kinetic coefficient,

$$\widetilde{\widetilde{\beta}}\left(\mathbf{n}\right) = \frac{T}{L^\alpha} \left[ \frac{\tau_{\alpha\beta}\left(\mathbf{n}\right)}{a_c\left(\mathbf{n}\right)} - \varepsilon a_c\left(\mathbf{n}\right) \frac{\left(c^\beta\left(\mu_{eq}, T\right) - c^\alpha\left(\mu_{eq}, T\right)\right)^2}{\left(D^\beta\right) \dfrac{\partial c^\beta\left(\mu^0, T\right)}{\partial \mu}} \left(M + F\right) \right].$$

$$(6.40)$$

To achieve vanishing interface kinetics in all directions we can choose, $\tau_{\alpha\beta}\left(\mathbf{n}\right)$ as $\tau_{\alpha\beta}^0 a_c^2\left(\mathbf{n}\right)$, where $\tau_{\alpha\beta}^0$ is derived from the expression in Eqn.(6.38). The case of anisotropy in kinetics would however require a more careful evaluation of the functions. Lastly, we would like to recall that a linearization of the grand potential around the equilibrium chemical potential was used for deriving the asymptotics. This is valid for small departures from equilibrium in phase transitions occurring at low undercoolings, where interface kinetics is absent and small. For certain situations at very high undercoolings and in the presence of strong kinetics, there might arise a situation where this linearization does not hold. This depends on the nature of the grand potentials and the magnitude of departure from equilibrium. The linearization is however only a simplification which can be easily relaxed resulting in the modification of the sharp interface limit for isotropic surface energies through the relation,

$$\Psi_\beta\left(T, \mu^0\right) - \Psi_\alpha\left(T, \mu^0\right) = \sigma\kappa + \tau_{\alpha\beta}V. \qquad (6.41)$$

The derivation of the deviation of the chemical potential however, depends on the nature of the grand potentials, where expansions of the grand potentials, until first or second order in the term $\mu^0 - \mu_{eq}$ might be necessary. The expression for the chemical potential at first order remains unchanged. However no general rule exists for estimating the validity of the linearization used in the asymptotics and the departure from equilibrium must be used to estimate the difference between the linearized and the original grand potential description, before reaching a conclusion.

## 6.2.5. Multicomponents and multiphases

One must note that although, the present study has been performed for the case of two-phase alloy solidification, this is not a limitation and the analysis can be easily generalized similar to earlier works [30]. For example: the anti-trapping current for the case of multi-phase, multi-component alloy solidification, where all the solid phases have zero diffusivities can be obtained by averaging each of the individual fluxes for each component $i$ given by,

$$\left(j_{at}^{\alpha \rightarrow l}\right)_i = -\frac{\pi \varepsilon}{4} \frac{g_\alpha\left(\phi_\alpha^0\right)\left(1 - h_\alpha\left(\phi_\alpha^0\right)\right)}{\sqrt{\phi_\alpha^0\left(1 - \phi_\alpha^0\right)}} \left(c_i^\beta\left(\boldsymbol{\mu}^0, T\right) - c_i^\alpha\left(\boldsymbol{\mu}^0, T\right)\right) \frac{\partial \phi_\alpha}{\partial t} \frac{\nabla \phi_\alpha}{|\nabla \phi_\alpha|},$$

and summing up all the fluxes projected along the normal to the liquid phase-field contour as,

$$\left(j_{at}\right)_i = \sum_{\alpha=1}^{N} \left(j_{at}^{\alpha \rightarrow l}\right)_i \left(-\frac{\nabla \phi_\alpha}{|\nabla \phi_\alpha|} \cdot \frac{\nabla \phi_l}{|\nabla \phi_l|}\right).$$

Other possibilities also exist, and the generalized normal vector $q_{\alpha l}$ can itself be used for the projection. Similarly, the relaxation constant for a vanishing interface kinetics for the phase transitions from $\alpha$ to $\beta$ for more than two components can be obtained as an extension of our present analysis, where the expression for the case of isotropic surface energies can be easily seen as a modification of Eqn.(6.38). We modify Eqn.(6.29) and derive the corresponding expression for two-phase multi-component system of $K$ independent components, by solving for the system of equations for each of the components of the vector $\boldsymbol{\mu}$, at first order in $\varepsilon$. The expression is given by,

$$\left\{ \widetilde{\mu}_i\big|^{\pm} - \mu_{eq}^i \right\} = \left\{ \mu_i^0 - \mu_{eq}^i \right\} + \varepsilon V \left( \tilde{M} + \tilde{F} \right) \left[ D_{ij}^{\beta} \frac{\partial c_j^{\beta}}{\partial \mu_i} \right] \left\{ c_i^{\beta} \left( \boldsymbol{\mu}^0, T \right) - c_i^{\alpha} \left( \boldsymbol{\mu}^0, T \right) \right\}$$

Note, in the above notation $[]$ represents a matrix while $\{\}$ represents a vector. Multiplying, throughout with $\left[ \dfrac{\partial \Psi^{\beta} \left( T, \boldsymbol{\mu}_{eq} \right)}{\partial \mu_i} - \dfrac{\partial \Psi^{\alpha} \left( T, \boldsymbol{\mu}_{eq} \right)}{\partial \mu_i} \right] = \left[ c_i^{\alpha} \left( \boldsymbol{\mu}_{eq}, T \right) - c_i^{\beta} \left( \boldsymbol{\mu}_{eq}, T \right) \right]$, we derive,

$$\left[ c_i^{\alpha} \left( \boldsymbol{\mu}_{eq}, T \right) - c_i^{\beta} \left( \boldsymbol{\mu}_{eq}, T \right) \right] \left\{ \widetilde{\mu}_i\big|^{\pm} - \mu_{eq}^i \right\} =$$

$$\left[ c_i^{\alpha} \left( \boldsymbol{\mu}_{eq}, T \right) - c_i^{\beta} \left( \boldsymbol{\mu}_{eq}, T \right) \right] \left\{ \mu_i^0 - \mu_{eq}^i \right\} +$$

$$\varepsilon V \left( \tilde{M} + \tilde{F} \right) \left[ c_i^{\alpha} \left( \boldsymbol{\mu}_{eq}, T \right) - c_i^{\beta} \left( \boldsymbol{\mu}_{eq}, T \right) \right] \left[ D_{ij}^{\beta} \frac{\partial c_j^{\beta}}{\partial \mu_i} \right] \left\{ c_i^{\beta} \left( \boldsymbol{\mu}^0, T \right) - c_i^{\alpha} \left( \boldsymbol{\mu}^0, T \right) \right\}.$$

The first term on the right hand side of the preceding equation is given by the sharp interface limit $\tau_{\alpha\beta} V$, and to derive vanishing kinetics, the left hand side of the equation should vanish, which gives the following relation for the case of isotropic free energies in the case of multi-component systems:

$$\tau_{\alpha\beta} =$$

$$\varepsilon \left[ c_i^{\beta} \left( \boldsymbol{\mu}_{eq}, T \right) - c_i^{\alpha} \left( \boldsymbol{\mu}_{eq}, T \right) \right]_{1 \times K} \left[ D_{ij}^{\beta} \frac{\partial c_i^{\beta} \left( \boldsymbol{\mu}^0, T \right)}{\partial \mu_j} \right]_{K \times K}^{-1} \times$$

$$\left\{ c_j^{\beta} \left( \boldsymbol{\mu}^0, T \right) - c_j^{\alpha} \left( \boldsymbol{\mu}^0, T \right) \right\}_{K \times 1} \times \left( \tilde{M} + \tilde{F} \right), \tag{6.42}$$

$K$ being the number of independent components in the system. A similar expression can be seen in the work by Kim et al.[54].

From the above discussion we have all the corrections that we need for performing quantitative simulations. The corrections to the Stefan condition at higher orders, that are the *interface stretching* and the *surface diffusion*, vanish when anti-symmetric functions are used to interpolate the phase diffusivities, $g_\alpha (\phi_\alpha)$ and $h_\alpha (\phi_\alpha)$ for the grand potentials which are applied from results derived in previous literature [30, 46].

## 6.3. Conclusions

We present a multi phase-field model based on the grand potential functional. This modification enables to effectively decouple the bulk and interface contributions which in turn allows to upscale the length of simulations. This formulation is consistent with existing quantitative phase-field models and places it in a common framework starting from a grand potential functional. We perform an asymptotic analysis of the derived model and obtain the thin-interface limit for the kinetic coefficient and an expression for the anti-trapping current for the special case of *double obstacle type* potentials. It is noteworthy to mention that, computationally, the obstacle type potentials are more efficient because the interface is finitely defined. Hence, computations of the gradient terms can be finitely limited to a fixed number of points in the interface. This computationally efficiency was offset until now because there was were no existing thin-interface limit and no expressions for the anti-trapping current for these type of potentials. This precluded the possibility of performing any quantitative simulations for the case of alloys. However, with the present thin interface asymptotics this can now be realized. With such modifications, quantitative simulations of multi-phase, multi-component systems at larger scales have become computationally feasible.

# Chapter 7

# Generalized construction of parabolic free energies

# 7.1. Introduction

Phase-field modeling has become a fairly versatile technique for the treatment of problems involving phase transitions. In particular, problems in solidification involving multi-component alloys, are fairly elegantly treated. A key ingredient while performing phase-field simulations for the case of phase-transformation in alloys are the description for free-energies of the respective phases. While thermodynamic databases such as CALPHAD provide the necessary information, it is often convenient to construct simpler descriptions in the region of interest, in order to perform computationally efficient simulations. With this motivation, we take a brief survey of the available solution models and the relevant simplified constructions that are possible while retrieving the correct physics. It is important to note that although, the free energies provided by the CALPHAD databases are Gibbs- free energies, the same can be used for phase-field simulations where the conditions are at constant pressure and constant system volume. In such cases, the Gibbs-free energies and the Helmholtz- free energies which are required for phase-field simulations differ only by an integration constant.

# 7.2. Solution models for binary alloys

The solution models, that are normally used for fitting to the various binary alloys, fall into three categories,

- Ideal solution models

- Regular solution models

- Sub-regular solution models

An ideal solution model can be characterized when, there exists no atomic interactions and the excess enthalpy of mixing and excess mixing entropy are zero. The corresponding expression can be exemplary written in the following form for a binary alloy as,

$$G^{\alpha} = c_A G_A^0 + c_B G_B^0 + RT\left(c_A \ln\left(c_A\right) + c_B \ln\left(c_B\right)\right),$$

where $G_A^0$ and $G_B^0$ are the reference free energies of the components $A, B$ with the crystal structure of phase $\alpha$. The next extension, is the case, where, there exists an interaction between the two atoms of different type and the corresponding interaction energy is denoted by the term $\Omega_{AB}$. However, the interaction is symmetric. The enthalpy of mixing is non-zero, but there exists no excess entropy of mixing. The regular binary solution model is given by,

$$G^\alpha = c_A G_A^0 + c_B G_B^0 + \Omega_{AB} c_A c_B + RT \left( c_A \ln\left(c_A\right) + c_B \ln\left(c_B\right) \right).$$

The final step in the generalization, is of course where the interaction parameter is not symmetric with respect to the two atomic species, but is a function of the concentration. This is sub-regular solution model written as,

$$G^\alpha = c_A G_A^0 + c_B G_B^0 + \left(c_A \Omega_{AB} + c_B \Omega_{BA}\right) c_A c_B +$$
$$RT \left( c_A \ln\left(c_A\right) + c_B \ln\left(c_B\right) \right).$$

In general all of the above models, can be treated as special cases of the *Redlich-Kister Polynomials*. While higher order descriptions are also possible, it is often desirable to limit these descriptions until the sub-regular type, such that the eventual construction of ternary and multi-component systems is less complex. However, in relation to phase-field models, one does not require, the information of the free energies in the whole range of concentrations, as the departure from equilibrium is in most cases in a very small range. In the following discussion, we look into the properties that are required, in order to derive a minimalistic construction of the free energy data while deriving the right physics.

## 7.3. Basic thermodynamics

When performing phase-field simulations, one of the most important requirements is the correct coupling of the phase-field evolution to the concentration equation, namely the relation between the driving force and the surface tension through the Gibbs-Thomson equation. For the case of alloys, it is derived in the following discussion, by equating the driving force for phase transformation(grand potential difference), to the capillary force given by,

$$\Delta\Psi = \tilde{\sigma}_{\alpha\beta}\kappa,$$

where $\Delta\Psi$ is the driving force for phase transition from $\alpha$ to $\beta$ written as, $\Psi_\beta - \Psi_\alpha$. $\tilde{\sigma}_{\alpha\beta}$ is the surface tension of the interface between the phases $\alpha$ and $\beta$. We can relate the difference in the grand potentials to the shift in the chemical potentials from equilibrium through a linear expansion of the grand potentials about the equilibrium chemical potentials giving,

$$\left(\frac{\partial\Psi^\beta}{\partial\mu} - \frac{\partial\Psi^\alpha}{\partial\mu}\right)\Delta\mu = \tilde{\sigma}_{\alpha\beta}\kappa,$$

where $\Delta\mu$ is $\mu - \mu_{eq}$, $\mu_{eq}$ being the equilibrium chemical potential. Using the thermodynamic relation $\dfrac{\partial\Psi}{\partial\mu} = -c$, we derive,

$$\left(c^\alpha - c^\beta\right)\Delta\mu = \tilde{\sigma}_{\alpha\beta}\kappa.$$

Next we expand $\mu$ about the equilibrium compositions of either phase, to derive $\Delta\mu = \dfrac{\partial\mu}{\partial c}\left(c^\beta - c^\beta_{eq}\right) = \dfrac{\partial^2 f^\beta}{\partial c^2}\Delta c^\beta$, which upon substitution in the Gibbs-Thomson condition derives,

$$\Delta c^\beta = \frac{\tilde{\sigma}_{\alpha\beta}\kappa}{\dfrac{\partial^2 f^\beta}{\partial c^2}\left(c^\alpha - c^\beta\right)}$$

From the above discussion, it is clear that the shift in the equilibrium concentrations varies inversely as the second derivative of the free energy with respect to the concentration and the magnitude of rejection of solute. Hence, in order to arrive at the correct Gibbs-Thomson effect the evaluation of $\dfrac{\partial^2 f^\beta}{\partial c^2}$ and the right solute rejection: $\left(c^\alpha - c^\beta\right)$, from the constructed free energies is important. To establish this, the simplest free energies one can construct are indeed, second order polynomials. In the following, we derive the methodology for the construction of free energy density descriptions for binary alloys, and then extend it for the case multi- component alloys.

## 7.4. Description of binary alloys with parabolic type free energies

We start the construction of free energies of the respective phases with the following type of expression for the free energies,

$$f^\alpha (T, c) = A^\alpha(T)c^2 + B^\alpha(T)c + E^\alpha(T),$$

where the coefficients $A^\alpha (T)$, $B^\alpha (T)$ and $E^\alpha (T)$ are functions of temperature T. Our aim is to fit a simplified form for the free energies utilizing the data obtained from the CALPHAD databases for the specific system. We can determine the terms $A^\alpha (T)$ as $\frac{\partial^2 f^\alpha}{\partial c^2}|_{c_{eq}} \equiv \frac{1}{V_m} \frac{\partial^2 G^\alpha}{\partial c^2}|_{c_{eq}}$, computed at the equilibrium concentration of the phase at the temperature T, where $G^\alpha(T, c)$ is the free energy function obtained from the CALPHAD database. Next we derive the chemical potential $\mu_{eq} = \frac{1}{V_m} \frac{\partial G^\alpha}{\partial c}|_{c_{eq}}$ from the database and compute $B^\alpha (T)$ by equating the first derivative of the constructed free energies to the chemical potential from the database giving,

$$B^\alpha (T) = \mu_{eq} - 2A^\alpha (T) c_{eq}.$$

The only term left out is $E (T)$, which is fitted by equating it to the grand potential at the concentration $c_{eq}$ given by,

$$E^\alpha (T) = \Psi_{eq} - A^\alpha (T) c_{eq}^2,$$

where $\Psi_{eq} = \frac{1}{V_m} (G^\alpha (T, c_{eq}) - \mu_{eq}c)$. With these equations we can adequately fit, all the coefficients in the constructed free energy at the given temperature T. For a non-isothermal description it is essential to derive the equations in the neighborhood of the temperature one is simulating and perform a fitting in the temperature space. In most cases, a linear temperature fit suffices.

### 7.4.1. Mobilities for diffusion

To derive the mobilities, we require the second derivatives of the free energies with respect to the concentration. This can be realized through the following diffusion equation for a binary alloy written as follows,

$$\frac{\partial c^\alpha}{\partial t} = \nabla \cdot (M^\alpha \nabla \mu),$$

where $M^\alpha$ is defined as $D\dfrac{\partial c^\alpha}{\partial \mu}$. From the free energy expression one can derive, the concentration as a function of the chemical potential as,

$$c^\alpha(\mu) = \frac{(\mu - B^\alpha(T))}{2A^\alpha(T)}.$$

Using the relation $c^\alpha(\mu)$, one can derive the relation $\dfrac{\partial c^\alpha}{\partial \mu} = \dfrac{1}{2A^\alpha(T)}$ and hence the mobility $M^\alpha$.

## 7.5. Extension to the case of ternary alloys(three components)

The extension of the parabolic free energy formulation for the case of ternary alloys is relatively straight forward. We write the free energies of the respective phases in the following form,

$$f^\alpha(c_A, c_B, c_C, T) = A^\alpha(T) c_A^2 + B^\alpha(T) c_B^2 + C^\alpha(T) c_C^2 +$$
$$O^\alpha(T) c_A + P^\alpha(T) c_B + Q^\alpha(T),$$

where $A^\alpha, B^\alpha, C^\alpha$ are the components and $c_A, c_B, c_C$ their respective concentrations. To determine the coefficients of the polynomial, we follow the same route as before for the binary alloy, by first writing the second derivatives with respect to composition as,

$$\frac{\partial^2 f^\alpha}{\partial c_A^2} = 2\left(A^\alpha\left(T\right) + C^\alpha\left(T\right)\right) \equiv \frac{1}{V_m}\frac{\partial^2 G^\alpha}{\partial c_A^2}$$

$$\frac{\partial^2 f^\alpha}{\partial c_B^2} = 2\left(B^\alpha\left(T\right) + C^\alpha\left(T\right)\right) \equiv \frac{1}{V_m}\frac{\partial^2 G^\alpha}{\partial c_B^2}$$

$$\frac{\partial^2 f^\alpha}{\partial c_A c_B} = \frac{\partial^2 f^\alpha}{\partial c_B c_A} = 2C^\alpha\left(T\right) \equiv \frac{1}{V_m}\frac{\partial^2 G^\alpha}{\partial c_A c_B}.$$

Through the above equations, we can fix the coefficients $A^\alpha\left(T\right), B^\alpha\left(T\right)$ and $C^\alpha\left(T\right)$. To derive the coefficients $O^\alpha\left(T\right)$ and $P^\alpha\left(T\right)$ we write the two independent equilibrium chemical potentials as,

$$\mu_A = 2A^\alpha\left(T\right)c_A - 2C^\alpha\left(T\right)c_C + O^\alpha\left(T\right) \equiv \frac{1}{V_m}\left(\frac{\partial G^\alpha}{\partial c_A}\right)_{c_B}$$

$$\mu_B = 2B^\alpha\left(T\right)c_B - 2C^\alpha\left(T\right)c_C + P^\alpha\left(T\right) \equiv \frac{1}{V_m}\left(\frac{\partial G^\alpha}{\partial c_B}\right)_{c_A}.$$

With the above equations all the terms in the polynomial can be fixed except the term $Q^\alpha\left(T\right)$, which is be determined, by equating the grand potential at the equilibrium concentrations at the given temperature, to the value obtained from the database given as $\Psi_{eq} = \frac{1}{V_m}\left(G^\alpha\left(c_A, c_B.c_C, T\right) - \mu_A c_A - \mu_B c_B\right)$. The coefficients $A^\alpha$ through $P^\alpha$ can be written as,

$$C^\alpha\left(T\right) = \frac{1}{2V_m}\frac{\partial^2 G^\alpha}{\partial c_A c_B}$$

$$A^\alpha\left(T\right) = \frac{1}{2V_m}\frac{\partial^2 G^\alpha}{\partial c_A^2} - C^\alpha\left(T\right)$$

$$B^\alpha\left(T\right) = \frac{1}{2V_m}\frac{\partial^2 G^\alpha}{\partial c_B^2} - C^\alpha\left(T\right)$$

$$O^\alpha\left(T\right) = \mu_A - 2A^\alpha\left(T\right)c_A + 2C^\alpha\left(T\right)c_C$$

$$P^\alpha\left(T\right) = \mu_B - 2B^\alpha\left(T\right)c_B + 2C^\alpha\left(T\right)c_C.$$

The concentrations $c_A\left(\mu_A,\mu_B,T\right)$ and $c_B\left(\mu_A,\mu_B,T\right)$ can be derived by inverting the expressions for the chemical potential simultaneously, and the corresponding expressions are,

$$c_A^\alpha\left(\mu_A,\mu_B,T\right)=\dfrac{\dfrac{\mu_A-O^\alpha\left(T\right)+2C^\alpha\left(T\right)}{2C^\alpha\left(T\right)}-\dfrac{\mu_B-P^\alpha\left(T\right)+2C^\alpha\left(T\right)}{2\left(B^\alpha\left(T\right)+C^\alpha\left(T\right)\right)}}{\dfrac{A^\alpha\left(T\right)+C^\alpha\left(T\right)}{C^\alpha\left(T\right)}-\dfrac{C^\alpha\left(T\right)}{B^\alpha\left(T\right)+C^\alpha\left(T\right)}}$$

$$c_B^\alpha\left(\mu_A,\mu_B,T\right)=\dfrac{\dfrac{\mu_A-O^\alpha\left(T\right)+2C^\alpha\left(T\right)}{2\left(A^\alpha\left(T\right)+C^\alpha\left(T\right)\right)}-\dfrac{\mu_B-P^\alpha\left(T\right)+2C^\alpha\left(T\right)}{2C^\alpha\left(T\right)}}{\dfrac{C^\alpha\left(T\right)}{A^\alpha\left(T\right)+C^\alpha\left(T\right)}-\dfrac{B^\alpha\left(T\right)+C^\alpha\left(T\right)}{C^\alpha\left(T\right)}}.$$

To derive the mobilities we need the matrix,

$$\begin{bmatrix} \dfrac{\partial c_A^\alpha}{\partial \mu_A} & \dfrac{\partial c_A^\alpha}{\partial \mu_B} \\[2ex] \dfrac{\partial c_B^\alpha}{\partial \mu_A} & \dfrac{\partial c_B^\alpha}{\partial \mu_B} \end{bmatrix},$$

which is inverse of the matrix,

$$\begin{bmatrix} \dfrac{\partial^2 f^\alpha}{\partial c_A^2} & \dfrac{\partial^2 f^\alpha}{\partial c_A \partial c_B} \\[2ex] \dfrac{\partial^2 f^\alpha}{\partial c_B \partial c_A} & \dfrac{\partial^2 f^\alpha}{\partial c_B^2} \end{bmatrix}.$$

The inverse matrix can be evaluated as follows,

$$\dfrac{1}{2\left[A^\alpha\left(T\right)B^\alpha\left(T\right)+B^\alpha\left(T\right)C^\alpha\left(T\right)+C^\alpha\left(T\right)A^\alpha\left(T\right)\right]}\times$$
$$\begin{bmatrix} \left(B^\alpha\left(T\right)+C^\alpha\left(T\right)\right) & -C^\alpha\left(T\right) \\ -C^\alpha\left(T\right) & \left(A^\alpha\left(T\right)+C^\alpha\left(T\right)\right) \end{bmatrix}$$

## 7.6. General description for the case of multicomponent alloys

To generalize the procedure for the determination of the free energies to multi-component systems we start with $K-1$ parabolas for each of the $K-1$ independent components which are chosen arbitrarily and write the general form as follows,

$$f^{\alpha}\left(\boldsymbol{c},T\right) = \sum_{i=1}^{K-1} A_i^{\alpha}\left(T\right) c_i^2 + \sum_{\substack{j=1,k=1 \\ j<k}}^{K-1} B_{jk}^{\alpha}\left(T\right) c_j c_k + \sum_{i=1}^{K-1} O_i^{\alpha}\left(T\right) c_i + Q^{\alpha}\left(T\right).$$

The coefficients $B_{jk}^{\alpha}\left(T\right)$ are determined by equating to terms of the matrix, $\dfrac{1}{V_m}\left[\dfrac{\partial}{\partial c_j}\left(\dfrac{\partial G^{\alpha}}{\partial c_k}\right)\right]_{j!\neq k}$ which are computed using the information of the free energies from the databases. Similarly coefficients $A_i^{\alpha}\left(T\right)$ are evaluated by using the relation, $A_i^{\alpha}\left(T\right) = \dfrac{1}{V_m}\left[\dfrac{\partial}{\partial c_i}\left(\dfrac{\partial G^{\alpha}}{\partial c_i}\right)\right]$. The remaining coefficients $O_i^{\alpha}\left(T\right)$ are derived by equating the chemical potentials using the relation, $O_i^{\alpha}\left(T\right) = \mu_i^d - 2A_i^{\alpha}\left(T\right)\left(c_i^{\alpha}\right)_d - \sum_{\substack{j=1 \\ j\neq i}} B_{ij}\left(T\right)\left(c_j^{\alpha}\right)_d$. The last term in the expansion is the coefficient $Q^{\alpha}\left(T\right)$ which is fixed by equating the grand potential to the value in the database, $\dfrac{1}{V_m}\left(G^{\alpha}\left(\boldsymbol{c},T\right) - \sum_{i=1}^{K-1}\mu_i^d\left(c_i^{\alpha}\right)_d\right)$, where we have denoted the chemical potentials and the concentrations, used for fitting the coefficients for the free energy as $\mu_i^d$ and $\left(c_i^{\alpha}\right)_d$ respectively. The subscript "$d$" is short for "determining values". The values of the phase concentrations as functions of the chemical potential are linearized around these chosen values of concentration and chemical potential. The mobility matrix is derived from the fitted free energy function by inverting the matrix given by $V_m\left[\dfrac{\partial}{\partial c_i^{\alpha}}\left(\dfrac{\partial G^{\alpha}}{\partial c_j}\right)\right]^{-1}$, which we denote as the matrix $\left[\dfrac{\partial c_i^{\alpha}}{\partial \mu_j}\right]$. The concentrations $c_i^{\alpha}\left(\boldsymbol{\mu},T\right)$ can be written as

$$\left\{c_i^{\alpha}\left(\boldsymbol{\mu},T\right)\right\} = \left\{\left(c_i^{\alpha}\right)_d\right\} + \left[\dfrac{\partial c_i^{\alpha}}{\partial \mu_j}\right]\left\{\mu_i - \mu_i^d\right\}.$$

## 7.7. Derivation of chemical potential

For the phase-field model using the grand potential formalism it is necessary to derive the chemical potential $\mu_i$, given the phase-field and the concentration fields at a given point. The expression for the chemical potential can be derived using the general expression for each of the concentration fields as, $c_i = \sum_{\alpha=1}^{N} c_i^{\alpha} h_{\alpha}(\phi)$. Inserting the phase concentrations $c_i^{\alpha}(\mu, T)$ as functions of $\mu$ we derive,

$$\{c_i\} = \sum_{\alpha=1}^{N} \{(c_i^{\alpha})\}_d \, h_{\alpha}(\phi) + \left[ \sum_{\alpha=1}^{N} h_{\alpha}(\phi) \frac{\partial c_i^{\alpha}}{\partial \mu_j} \right] \{\mu_i - \mu_i^d\},$$

which can be re-arranged to derive the components of the vector $\mu$ as,

$$\{\mu_i - \mu_i^d\} = \left[ \sum_{\alpha=1}^{N} h_{\alpha}(\phi) \frac{\partial c_i^{\alpha}}{\partial \mu_j} \right]^{-1} \left\{ c_i - \sum_{\alpha=1}^{N} (c_i^{\alpha})_d \, h_{\alpha}(\phi) \right\}.$$

Note the calculation of the vector components is possible in this fashion because the components of the matrix $\left[ \dfrac{\partial c_i}{\partial \mu_j} \right]$ are independent of $\mu$, being functions only of temperature $T$ for this special method of construction of free energies.

## 7.8. Application for the case of AlCuAg ternary alloy

We obtain the information about the free energies of the phases $\alpha$-FCC, $\theta$ and the $\gamma$-Hcp phases from the database created by Witusiewicz et al [137, 138]. The energies are fitted around the eutectic temperature $T_E$ using parabolic free energy forms. The respective coefficients are then linearly fitted as functions of temperature. (Details are given in the following chapter).

Utilizing the free energies, the grand potential formulation is used to derive the evolution equations for the phase-field and concentration or chemical potential field. Exemplary structures are portrayed, showing the growth of lamellae in a given temperature field. Contrary to binary eutectics, we find the growth front to be highly non-planar and the velocities of the three phases to be very different. The occurrence of coupled growth is not only a function of the undercooling but also of the lamellar spacing and the imposed velocity. We find that the possibility of coupled growth increases with decreased velocities and increased temperature gradients.

## 7.9. Conclusions

We present a simplified construction of free energies for phase-field simulations using information from CALPHAD databases. The construction allows for computationally simpler functions to be used, while retaining the physics of the problem. One must note, that the fitting methods adopted here, involves deriving the free energy data by linearizing about chosen concentrations. Depending on the deviations from equilibrium, one must adapt the fitting procedure, to avoid large deviations from reality. In general, simulations mimicking experimental conditions, result in small deviations of the phase concentrations from their equilibrium values. This allows one to use the described procedure, to fit the energies around the equilibrium compositions of the co-existing phases. Far from equilibrium, would require dynamic adaptation of the fitted data derived out of the local shifts in the equilibrium concentrations. Considering the simplified nature of the fitting procedure, it is certainly possible to perform it even during run-time when required.

# Chapter 8

# Validation

In this chapter we present test cases, where the modified model based on the grand potential formulation, is utilized for the simulation of dendritic and eutectic microstructures. For the case of dendritic growth, we consider the Al-Cu alloy, while for ternary eutectic growth, we utilize the model ternary eutectic system, created before for comparison with theoretical analysis of coupled growth. Additionally we model the Al-Cu-Ag alloy for some preliminary studies. We show exemplary structures, found generally during solidification in each of the forthcoming sections.

## 8.1. Dendritic growth

The phase stability regions in the Al-Cu phase diagram containing the $\alpha$-Al and the liquid phase, respectively, are confined by nearly linear phase stability lines corresponding to slopes of 45.3K/wt% and 2.6K/wt%, for solidus and liquidus respectively. The phase diagram was modeled using the ideal solution formulation, where the free energy density, $f^\alpha(T,c)$ with $c=c_{Al}$ is described as:

$$f^\alpha(T,c) = f^* \left( cL^\alpha_{Al} \frac{(T - T^\alpha_{Al})}{T^\alpha_{Al}} + (1-c)L^\alpha_{Cu} \frac{(T - T^\alpha_{Cu})}{T^\alpha_{Cu}} + \right.$$
$$\left. T\left(c\ln c + (1-c)\ln(1-c)\right) \right) \qquad (8.1)$$

where the reference temperature for nondimensionalizing is chosen to be the melting temperature of pure Al. The parameters $L^\alpha_i$ and $T^\alpha_i$ are tabulated in Table 8.1, The Gibbs-Thomson coefficient $\Gamma_{\alpha l}$ is 2.48 $10^{-7}$K/m [38],

**Table 8.1.:** Free energy parameters

| $L^\alpha_i$ | Cu | Al | | $T^\alpha_i$ | Cu | Al |
|---|---|---|---|---|---|---|
| $\alpha$ | 8.45 | 5.30 | | $\alpha$ | 0.42273 | 1.0 |
| liquid | 0.0 | 0.0 | | liquid | X | X |

and energy scale $f^*$ is calculated from the Gibbs-Thomson coefficient and the given surface tension. The modeled phase diagram is plotted in Fig.8.1. With this choice of free energies, the functions $c^{\alpha/l}(\mu, T)$ can be

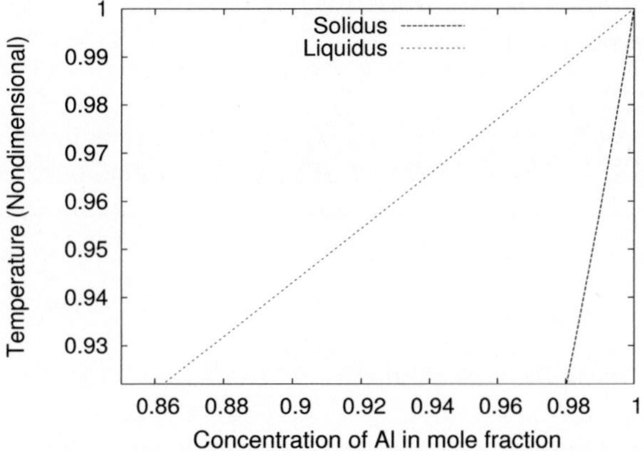


**Figure 8.1.:** Phase stability region between the $\alpha-$ solid and liquid modeled using ideal free energies

determined in the following manner:

$$c^{\alpha}\left(\mu, T\right) = \frac{\exp\left[\dfrac{\mu - \left(L^{\alpha}_{Cu}\dfrac{(T - T^{\alpha}_{Cu})}{T^{\alpha}_{Cu}} - L^{\alpha}_{Al}\dfrac{(T - T^{\alpha}_{Al})}{T^{\alpha}_{Al}}\right)}{f^{*}T}\right]}{1 + \exp\left[\dfrac{\mu - \left(L^{\alpha}_{Cu}\dfrac{(T - T^{\alpha}_{Cu})}{T^{\alpha}_{Cu}} - L^{\alpha}_{Al}\dfrac{(T - T^{\alpha}_{Al})}{T^{\alpha}_{Al}}\right)}{f^{*}T}\right]}$$

$$c^{l}\left(\mu, T\right) = \frac{\exp\left[\dfrac{\mu}{f^{*}T}\right]}{1 + \exp\left[\dfrac{\mu}{f^{*}T}\right]}.$$

Notice that we have written $c^{\alpha}\left(\mu, T\right)$ and $c^{l}\left(\mu, T\right)$ as functions of a unique $\mu$, as required by the grand potential formulation.

## 8.1.1. Kinetics of diffusion and phase transformations

The diffusion coefficients were set to $D^\alpha = 3 \cdot 10^{-13} \mathrm{m}^2/\mathrm{s}$ in the solid phase and $D^l = 3 \cdot 10^{-9} \mathrm{m}^2/\mathrm{s}$ in the liquid. The anti-trapping current is defined as a flux from the solid to the liquid side across the $\alpha l$ interface as,

$$(j_{at}) = -\frac{\pi \varepsilon}{4} \frac{h_\alpha\left(\phi\right)\left(1 - h_\alpha\left(\phi\right)\right)}{\sqrt{\phi_\alpha^0\left(1 - \phi_\alpha^0\right)}} \left(c^l\left(\mu, T\right) - c^\alpha\left(\mu, T\right)\right) \frac{\partial \phi_\alpha}{\partial t} \frac{\nabla \phi_\alpha}{|\nabla \phi_\alpha|},$$

where $\phi_\alpha^0$ is the leading order solution of the phase-field equation, which is also the equilibrium phase-field profile. With the anti-trapping current, the diffusion equation is modified as follows:

$$\frac{\partial c}{\partial t} = \nabla \cdot \left(M\left(\phi\right) \nabla \mu - (j_{at})\right).$$

The kinetics of phase transformation in experiments at the micro-scale, occurs on a time scale that is orders of magnitude larger than the one given by the atomistic relaxation. Hence, for such cases the time-scale of interface relaxation is irrelevant. Therefore, it is desirable to have infinite mobility for phase-field evolution, which implies that the response of the phase-field, to a change in the coupled field is instantaneous. To achieve this in the phase-field methodology, a thin interface analysis needs to be performed, as demonstrated first by Karma [46]. We use such an analysis for the case of an obstacle potential presented in earlier chapters and in [18]. The results of the analysis give the choice of the relaxation time constant $\omega$ for the case of binary alloy as,

$$\omega = \varepsilon \frac{\left(c^l\left(\mu_{eq}, T\right) - c^\alpha\left(\mu_{eq}, T\right)\right)\left(c^l\left(\mu^0, T\right) - c^\alpha\left(\mu^0, T\right)\right)}{T\left(D^l\right) \dfrac{\partial c^l\left(\mu^0.T\right)}{\partial \mu}} \left(\tilde{M} + \tilde{F}\right)$$

$$(8.2)$$

With this choice, we obtain $\overline{\beta} = \dfrac{1}{\mu_{int}^\alpha} = 0$, where $\mu_{int}^\alpha$ is the kinetic coefficient of the $\alpha, l$ interface. $\tilde{M}$ and $\tilde{F}$ are solvability integrals depending on the choice of the potential $w\left(\phi\right)$ and on the interpolation functions

**Table 8.2.:** Values of the solvability integrals for the employed interpolation polynomials

| Potential | $\tilde{M}$ | $\tilde{F}$ |
|---|---|---|
| $h_\alpha\left(\phi\right) = \phi_\alpha^2 \left(3 - 2\phi_\alpha\right)$ | 0.063828 | 0.158741 |
| $h_\alpha\left(\phi\right) = \phi_\alpha^3 \left(10 - 15\phi_\alpha + 6\phi_\alpha^2\right)$ | 0.052935 | 0.129288 |

$h_\alpha\left(\phi\right)$. For the obstacle potential, the values of the solvability integrals, corresponding to the two interpolation polynomials in use, are listed in Table 8.2. $\mu^0$ for the case of binary alloys denotes the macroscopic chemical potential at the interface, in the sharp interface limit. The kinetic coefficient can be computed both for finite and infinite phase-field interface mobility. For our computations, we set the vanishing interface kinetics in all directions. Our simulations are performed with smooth cubic anisotropy of the form,

$$a_c\left(q_{\alpha\beta}\right) = 1 \pm \delta_{\alpha\beta}\left(3 - 4\frac{|q_{\alpha\beta}|_4^4}{|q_{\alpha\beta}|^4}\right),$$

where $|q_{\alpha\beta}|_4^4 = \sum_i^d (q_{\alpha\beta})_i^4$ and $|q_{\alpha\beta}|^4 = \left[\sum_{i=1}^d (q_{\alpha\beta})_i^2\right]^2$, $d$ being the number of dimensions. $\delta_{\alpha\beta}$ is the strength of the anisotropy, which is set to 0.0097 for the chosen alloy of Al-4wt%Cu [67]. To have vanishing interface kinetics in all directions, we employ the strategy illustrated in the earlier chapter on *Grand potential formulation and asymptotics*, where we utilize the expression for the kinetic coefficient given by Eqn. (6.40) and set the relaxation constant as a function of the normal vector $q_{\alpha\beta}$ given by,

$$\omega\left(q_{\alpha\beta}\right) = \omega_0 a_c^2\left(q_{\alpha\beta}\right),$$

where $\omega_0$ is computed using the relation in Eqn. (8.2). This gives the kinetic coefficient as a function of the normal vector $\mathbf{n} = \dfrac{q_{\alpha\beta}}{|q_{\alpha\beta}|}$ as,

$$\widetilde{\beta}\left(\mathbf{n}\right) = \frac{T}{L^\alpha} a_c\left(\mathbf{n}\right)\left[\omega_0 - \varepsilon\frac{\left(c^\beta\left(\mu_{eq}, T\right) - c^\alpha\left(\mu_{eq}, T\right)\right)^2}{T\left(D^\beta\right)\dfrac{\partial c^\beta\left(\mu^0, T\right)}{\partial\mu}}\left(M + F\right)\right].$$

The term inside the brackets in the preceding equation vanishes by the choice of $\omega_0$ and hence renders the effective kinetic coefficient, zero in all directions. The same technique is adopted in the work by [50].

The simulation setup involves a free dendrite growing into an uniformly undercooled melt. Utilizing the symmetry of the surface energy anisotropy, we simulate one quadrant of the dendrite. The first benchmark involves the proof of invariance of the dendrite tip velocities with varying interface widths. For this, we set the nondimensional bulk temperature at T=0.9843, where the melting temperature of the chosen alloy composition Al-1.732At%, is T=0.99. Fig. 8.2 plots the dendrite tip velocities upon

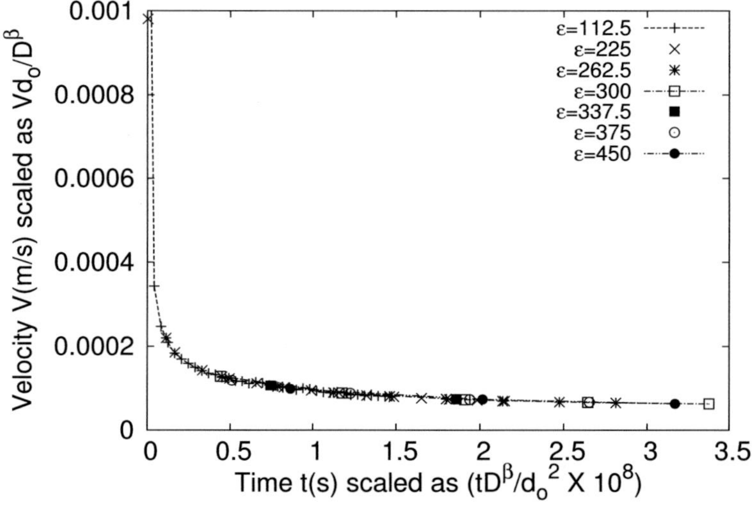


**Figure 8.2.:** Plot of the dendrite tip velocities simulated at a temperature of $T = 0.9843$. We selectively plot points corresponding to a simulation, to show the convergence of the velocities. We span a range where $\varepsilon$ varies by a factor 4 and achieve convergence in the velocities. The simulation with the $\varepsilon = 112.5$ has run the least in nondimensional time $(1.5\ 10^8)$, but long enough to confirm convergence of the velocities.

change in the interface widths which confirms our calculations, that their exists a range in interface widths for which the interface velocities are invariant. Above the maximum considered $\varepsilon$, the interface becomes unstable and the asymptotics seems to breakdown and we supposes this occurs because errors of order $O\left(\varepsilon^2\right)$ become appreciable. Fig. 8.3, displays the chemical potential along a linear section at the dendrite tip in the growing direction, along with the equilibrium chemical potential and the theoretical chemical potential derived from the Gibbs-Thomson condition considering only the effect of curvature. The results show good agreement, confirming

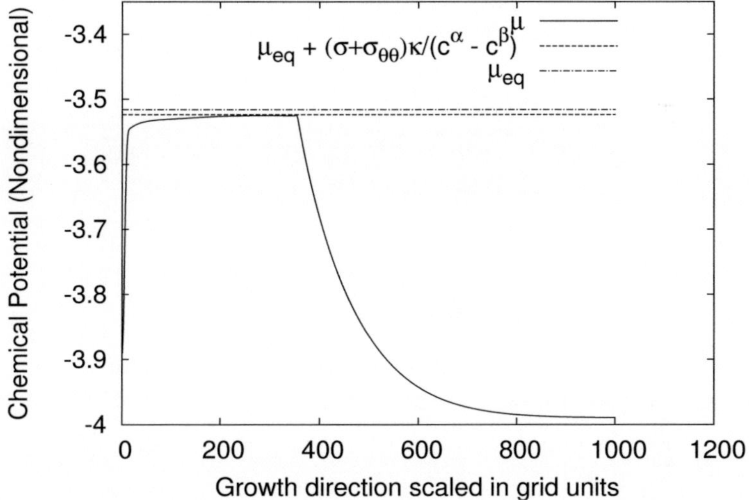


**Figure 8.3.:** Chemical potential plot along a linear section at the dendrite tip in the growing direction, superimposed with the lines showing the equilibrium chemical potential and the theoretically predicted chemical potential obtained by considering the shift because of Gibbs-Thomson effect due to curvature. The curvature used in the calculation, is measure at the dendrite tip from the simulation. $\sigma_{\theta\theta}$ represent the second derivative of the surface tension as a function of the polar angle, and the sum $\sigma + \sigma_{\theta\theta}$ represents the stiffness of the interface.

the asymptotic expressions and the applicability of the developed model. The contours of the chemical potential and the phase-profiles in a section of the domain showing the growing dendrite is portrayed in Fig. 8.4.

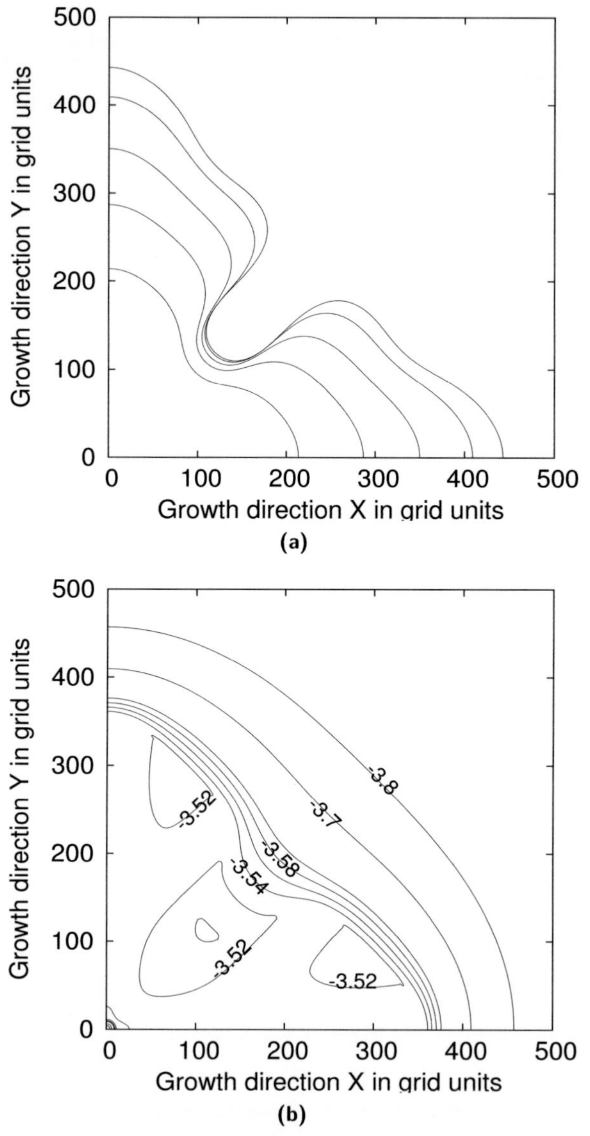



**Figure 8.4.:** Isolevel $\phi_\alpha = 0.5$ denoting the binary interface between the solid and the liquid at various times is shown in (a), while the contours of the chemical potential at a particular instant during evolution is displayed in (b). Values of some of the contours of the respective nondimensionalized chemical potential are superimposed on the plot. The simulations correspond to the case when $\varepsilon = 300$.

## 8.1.2. Comparison with the LGK

In order to verify our model, we test the dendrite tip and velocities against the analytical LGK theory [66]. An important parameter required for the matching, is the calculation of $\sigma^*$ which is the stability parameter used for determining the velocity and radius at the dendrite tip. While $\sigma^*$ is expected to be in the order of 0.025 in three-dimensional systems [66], realistic physical values vary with the anisotropy of the solid-liquid interface. Two possibilities exist to determine the value of $\sigma^*$. The first is to derive it from micro-solvability theory [10]. The second is to employ techniques such as phase field simulations, where the dendrite tip radii and the corresponding velocities are a result of the dynamic minimization of the grand potential difference, and the $\sigma^*$ is an output of the simulation [49, 50]. We adopt the second technique, and compute the value of $\sigma^*$ as $\dfrac{2D^l d_0}{VR^2}$ from the steady state tip velocity $V$ and tip radius $R$ from the simulation. Here $d_0$ is the capillary length, defined as $\dfrac{\Gamma}{m\,(\Delta C)}$, where $\Gamma$ is the Gibbs-Thomson coefficient, $\Delta C$ the magnitude of solute rejection at the interface, and $m$ the slope of the liquidus line. We obtain a value which gives a best fit for all considered undercoolings. The stability parameter $\sigma^*$ was computed as 0.169 in 2D simulations, as an average value from the undercoolings considered. The comparison is plotted in Fig.8.5. The simulations at each

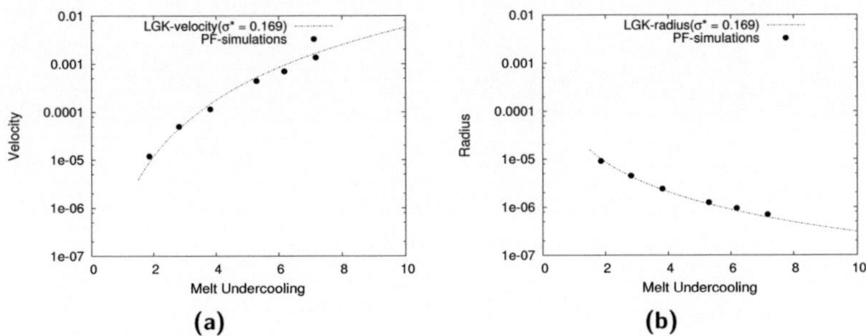


**Figure 8.5.:** Comparison of the velocity and radius at the dendrite tip as a function of the bulk undercooling.

undercooling was performed with interface widths which are in the range, where the results are invariant of this parameter. One must note that while the LGK theory gives good predictions, there can often arise a case where there is a variation. These arise because of the mismatch between the assumptions in the dendritic shape made in the LGK analysis and those occurring in the simulations. In the next section, we simulate some exemplary structures during solidification for this alloy.

### 8.1.3. Exemplary structures

Equi-axed dendritic growth occurs when the undercooling in the melt is high enough such that there exists no directionality in the local thermal gradients. Fig.8.6 shows dendrites at different undercoolings. Cellular

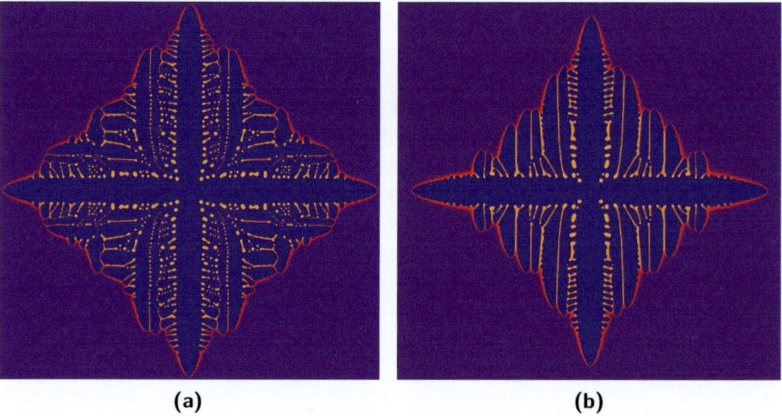

(a)                                                 (b)

**Figure 8.6.:** Equiaxed dendrites simulated at undercooling of 30K in (a) and 28K in (b) an alloy composition of 0.017382 at% Cu.

growth into a undercooled melt was simulated for two situations. In the first case shown in, Fig. 8.7a, the orientation of anisotropy is aligned in the direction of growth, while in Fig. 8.7b, the dendrites are rotated 30 degrees with respect to the growth direction. In 8.7c, the simulations are performed with a temperature gradient in the growth direction of value $6000K/m$ shifted, with a velocity of $0.05m/s$ along the growth axis.

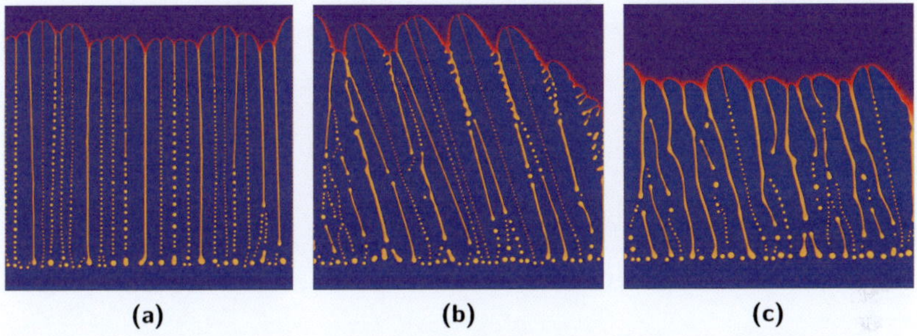


**Figure 8.7.:** Cellular growth structures in 2D. In (a) the orientation of the anisotropy is aligned with the growth direction while in (b) the crystal orientation is rotated 30 degrees with respect to the growth direction, and in (c), the simulation is performed with a temperature gradient aligned in the growth direction.

## 8.2. Eutectic growth

### 8.2.1. Comparison with theoretical expressions

We use the model ternary eutectic system used for the investigation in earlier chapters for comparison with analytical expressions derived according to theoretical calculations of the Jackson-Hunt type. The phase concentrations as functions of the chemical potential required for the construction of the grand- potentials are given by,

$$c_i^\alpha\left(\boldsymbol{\mu}, T\right) = \frac{\exp\left(\mu_i/T - \left(L_i^\alpha \dfrac{(T - T_i^\alpha)}{TT_i^\alpha} - L_C \dfrac{(T - T_C^\alpha)}{TT_C^\alpha}\right)\right)}{1 + \sum_{j=1}^{K-1}\exp\left(\mu_j/T - \left(L_j^\alpha \dfrac{(T - T_j^\alpha)}{TT_j^\alpha} - L_C \dfrac{(T - T_C^\alpha)}{TT_C^\alpha}\right)\right)}$$

(8.3)

Utilizing the construction, we verify the equilibrium properties, by measuring the triple point angles for a stationary solidification interface as shown in, Fig.8.8. It is noteworthy however, that the angles are now retrieved,

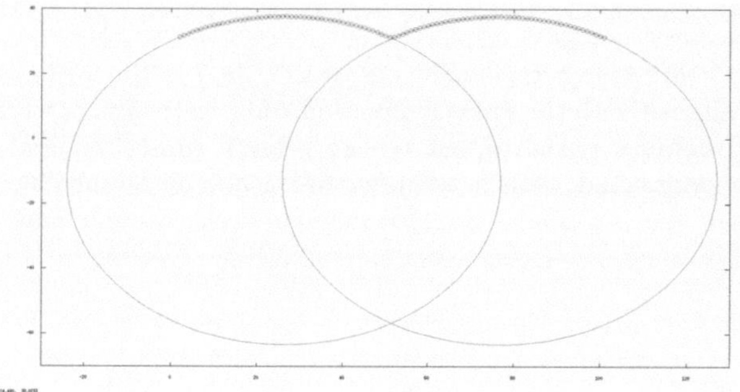

**Figure 8.8.:** Triple junction at the critical undercooling where the solidification front is stationary. The whole binary junction of either solid-liquid interfaces are fitted with circles and the angle at the triple point is measured between the tangents to the circles at the intersection point of the circles

just by setting the required surface energies of the respective interfaces as the simulation parameter $\gamma_{\alpha\beta}T$, unlike the calibration required with the free-energy model. Also, the third phase contribution is markedly reduced and can be completely removed with a value of $\gamma_{\alpha\beta\delta} = (10 - 15)\gamma_{\alpha\beta}$. This was not possible in the case of the model with the free energies, where much higher values where required, resulting in unwanted modification in the area around the triple point. This is illustrated by the individual phase-profiles plotted in Figs.8.9 and 8.10  In addition, we compare the $\alpha\beta\gamma$ and

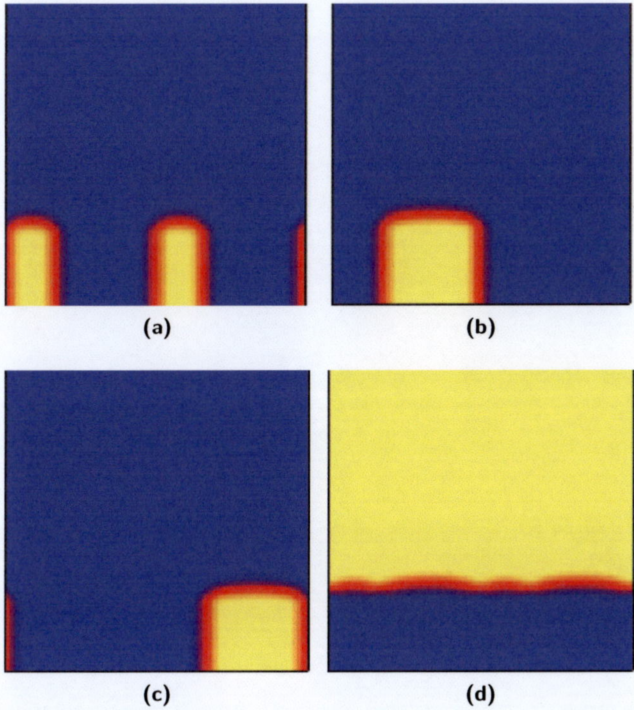

**Figure 8.9.:** Individual phase-profiles obtained from simulations with the grand potential model depicting no third phase adsorption at any of the interfaces, with a value of the higher order potential $\gamma_{\alpha\beta\delta} = 10\gamma_{\alpha\beta}$. The red border, plots the contour of the phase-field from 0 to 1.

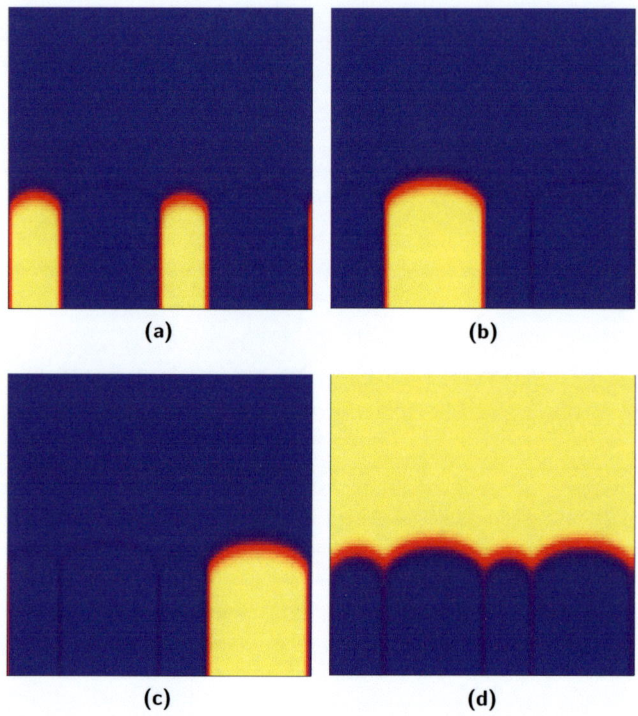

Figure 8.10.: Corresponding phase-profiles obtained from simulations with the model based on a free energy functional depicting third phase adsorption at all of the interfaces, with a value of the higher order potential $\gamma_{\alpha\beta\delta} = 10\gamma_{\alpha\beta}$. Higher, values of $\gamma_{\alpha\beta\delta}$ results in the distortion of the triple-point regions.

the $\alpha\beta\alpha\gamma$ configurations with the theoretical expressions for undercooling as functions of spacing at given velocities. We achieve similar agreement as before with the the free energy model as shown in, Fig.8.11, but this time

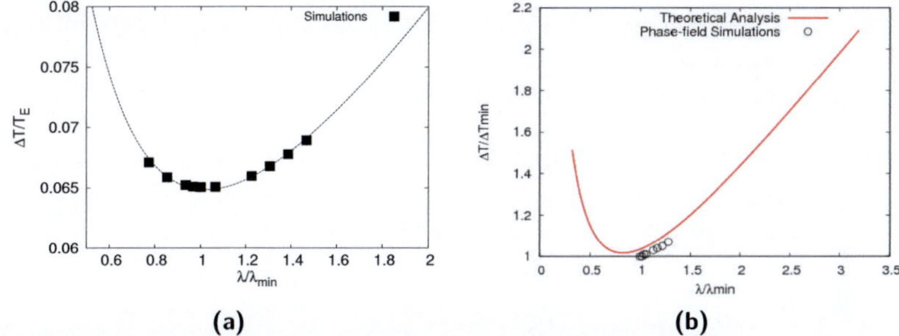


**Figure 8.11.:** Comparison of analytical theories with simulations for two configurations $\alpha\beta\gamma$ in (a) and $\alpha\beta\alpha\gamma$ in (b). Figure(a) shows the fit of the simulation points with the function $\Delta T = \dfrac{\Delta T_{min}}{2} \left( \dfrac{\lambda}{\lambda_{min}} + \dfrac{\lambda_{min}}{\lambda} \right)$, which gives $\Delta T_{min} = 0.0612$ for $(\Delta T_{JH} = 0.0648)$ with an error of 6 %, and similarly an error in $\lambda_{min} = 1.0052(\lambda_{JH} = 1.02)$ of 1.8 %.

the interface kinetics was removed, using the derived expressions for the kinetic coefficient obtained from the asymptotic analysis. The expression of the relaxation coefficient, to achieve vanishing interface kinetics for the case of ternary eutectics is derived through an extension of the expressions for the case of binary alloys, as given in Eqn. (6.42) written as:

$$\omega_{\alpha l} = \frac{\tilde{M} + \tilde{F}}{T} \left[ \frac{\left( c_A^l\left(\boldsymbol{\mu}, T\right) - c_A^\alpha\left(\boldsymbol{\mu}, T\right) \right)^2}{D \dfrac{\partial c_A^l}{\partial \mu_A}} + \right.$$

$$\left. \frac{2\left( c_A^l\left(\boldsymbol{\mu}, T\right) - c_A^\alpha\left(\boldsymbol{\mu}, T\right) \right)\left( c_B^l\left(\boldsymbol{\mu}, T\right) - c_B^\alpha\left(\boldsymbol{\mu}, T\right) \right)}{D \dfrac{\partial c_A^l}{\partial \mu_B}} + \frac{\left( c_B^l\left(\boldsymbol{\mu}, T\right) - c_B^\alpha\left(\boldsymbol{\mu}, T\right) \right)^2}{D \dfrac{\partial c_B^l}{\partial \mu_B}} \right],$$

where $\tilde{M}$ and $\tilde{F}$ are the solvability integrals derived in the earlier chapters, and tabulated in Table. 8.2, while the diffusivities in the liquid are assumed to be the same for both components A and B. For the case of $\alpha\beta\alpha\gamma$ the average undercoolings are lower than the predicted one. This is also what we achieved with the model derived from the free energy functional. The reason for the deviation can be explained through the observation, that we do not have a planar growth front, as is the assumption in the theoretical analysis. This is also coupled with the fact that the front undercoolings of the different lamellae are not the same which implies that the assumption of an isothermal growth front is not exact.

In addition, for the simulation of more phases it was essential to construct right interpolation functions. For this we corrected our third order interpolation polynomial in the following manner:

$$h_\alpha\left(\phi\right) = \phi_\alpha^2\left(3 - 2\phi_\alpha\right) + 2\phi_\alpha \sum_{\substack{\beta\neq\alpha \\ \gamma\neq\alpha}}^{N} \phi_\beta\phi_\gamma.$$

Other corrected polynomials are listed in the appendix C.


## 8.2.2. Effect of solid-solid anisotropy

The model system is then used to investigate the role of solid-solid anisotropy in oscillatory mode selection. For this we selected two configurations $\alpha\beta\gamma$ and $\alpha\beta\alpha\gamma$. From the analysis performed in the chapter on the ternary eutectics we used symmetry arguments to characterize the different oscillatory modes. In the presence of solid-solid anisotropy it is possible to manipulate the symmetry elements in the configuration, and our aim lies in relating the symmetry elements in the simulated microstructures, to those of the initial configuration, modified by solid-solid anisotropy. We choose the different permutations that are possible in modifying the symmetries. For instance, in the $\alpha\beta\gamma$ there exist three possibilities, where either one, two or all three solid-solid interfaces are anisotropic, while in the case of $\alpha\beta\alpha\gamma$, only two possibilities exist, which is either the $\alpha\beta$ or the $\alpha\gamma$ is anisotropic or both. Each of the permutations possesses different symmetry elements, where we impose smooth-cubic anisotropy oriented in the growth direction for the intended solid-solid interfaces. For the case of $\alpha\beta\gamma$, the simulated modes are listed in Fig.8.12. The symmetry

**Figure 8.12.:** Possible oscillatory modes achieved for the different combinations of anisotropic solid-solid interfaces. In (a)($\lambda = 115$) and (c)($\lambda = 120$) two interfaces are anisotropic while in (b)($\lambda = 120$), all three interfaces are anisotropic. (d) shows the characteristic mode achieved in (c) in enlarged form

elements in the configuration are reproduced in the oscillatory modes and we have all the possibilities as in the case of the isotropic surface energies. However, we have certain modifications such as in Figs. 8.12c and 8.12d, where the configuration with the absence of any symmetry plane has a stacking of the three respective phases along a plane which is tilted with respect to the growth axis. However, if the symmetry axes containing the anisotropic interface remains aligned with the growth direction, the configurations with the respective symmetry elements are retrieved. In the case of the configuration $\alpha\beta\alpha\gamma$, the possibilities are plotted in Fig.8.13. As was discussed before, there exist two ways to manipulate the symmetry elements of the configuration $\alpha\beta\alpha\gamma$ through the solid-solid anisotropies. While the 2-$\lambda$-O mode is retrieved in every constructed configuration, which can be reasoned based on the loss of the symmetry plane passing through the $\alpha$ phase and hence the resultant oscillatory mode shares the same symmetry elements as that of the starting configuration. However, we observe a modified mode in the presence of solid-solid anisotropy as in Fig.8.13a.

## 8.2.3. Ternary eutectic AlCuAg

**Free Energies and surface data**

We construct free energies using information from the CALPHAD data base for the Al-Cu-Ag system provided by [137, 138]. The free energies are fitted to simplified parabolic type free energies using the methodology described in the preceding chapter, around the eutectic temperature. The coefficients of the polynomial are fitted linearly with respect to temperature around the eutectic point.

The phases at equilibrium are FCC-$\alpha$, $Al_2Cu$-$\theta$ and (HCP)-$Ag_2Al$ $\gamma$ phase. With the thermodynamic information in the database, the volume fractions of the respective solid phases at equilibrium are $\eta_\alpha = 0.55$, $\eta_\theta = 0.29$ and $\eta_\gamma = 0.16$.

To derive information about the surface energies, we utilize experimental data provided for the measured equilibrium angles given in [41, 53] and we assume the surface energies of the $\alpha-$ liquid interface as 0.3 $J/m^2$. This fixes the interfacial energies of all the interfaces.

**Figure 8.13.:** Characteristic modes achieved for the different permutations of the anisotropic solid-solid interfaces. In (a) and(c) one interface is ansitropic while in (b) both possible solid-solid interfaces are anisotropic.

## Simulations

Preliminary structures are simulated in 2D. We find that $\alpha\beta\alpha\gamma$ configuration is one of the stable configurations. Noteworthy, is that the growth interface is not planar, and the different phase interfaces are not isothermal. The structures are grown at very low speeds of $(2 \ \mu m/s)$ used in solidification, with a temperature gradient of $13.33 \ K/mm$. Fig.8.14 shows the simulated structures and the undercooling as a function of the spacing, where the undercooling is averaged over the entire solid-liquid interface.

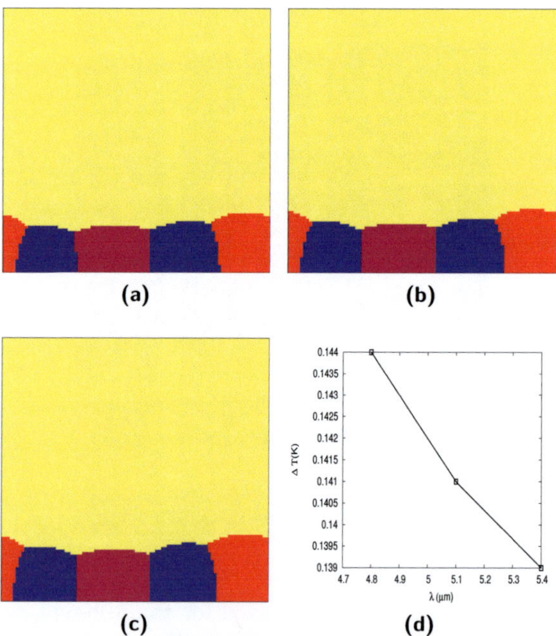

**Figure 8.14.:** Preliminary simulations of Al-Cu-Ag ternary eutectic alloy, showing proof of concept for the generalized construction of parabolic free energies for multi-component systems. The figure shows the $\alpha\beta\alpha\gamma$ configuration at three different lamella spacings in (a) $4.8\mu m$,(b) $5.1\mu m$ and in (c) $5.4\mu m$. The front undercoolings after 0.35s of solidification time, are plotted in (d).

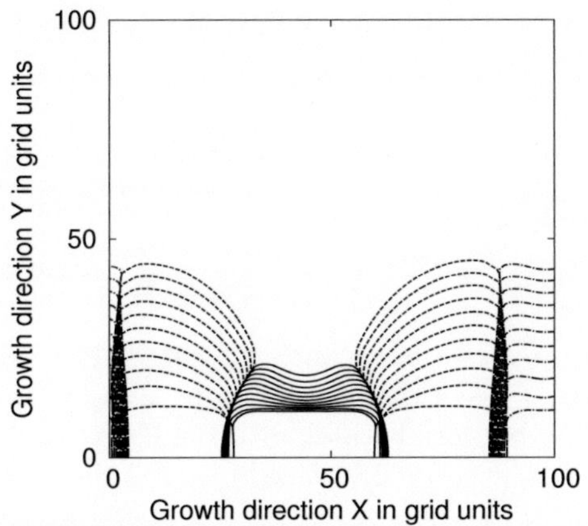

**Figure 8.15.:** Phase-profiles of the three solid phases at various instances during growth. From the left the phases are $\alpha(FCC)$, $\gamma(HCP)$ and $\theta$ respectively.

# 8.3. Free energy functional vs Grand potential functional

In this section, we present a short discussion on the range of applicability of the models derived from the grand potential and the free energy functional. The comparison is with respect to the computational efficiency in simulating a range of undercoolings for both models and the asymptotics. To start with, we write the evolution equation for the phase-field variables for a system of two phases and two components, starting from the free energy functional,

$$\tau_{\alpha\beta}\varepsilon\frac{\partial\phi_\alpha}{\partial t} = T\gamma_{\alpha\beta}\varepsilon\frac{\partial^2\phi_\alpha}{\partial x^2} - \frac{T\gamma_{\alpha\beta}}{\varepsilon}\frac{16}{2\pi^2}\left(1 - 2\phi_\alpha\right) - \frac{1}{2}\frac{df}{d\phi_\alpha} + \frac{1}{2}\frac{\partial f}{\partial c}\frac{dc}{d\phi_\alpha}.$$

Noting, that for this formulation based on the free energy functional, $\mu = \dfrac{\partial f}{\partial c}$, the equation can be manipulated as,

$$\tau_{\alpha\beta}\varepsilon^2\frac{\partial\phi_\alpha}{\partial t} =$$

$$T\gamma_{\alpha\beta}\varepsilon^2\frac{\partial^2\phi_\alpha}{\partial x^2} - \left\{T\gamma_{\alpha\beta}\frac{16}{2\pi^2}\left(1 - 2\phi_\alpha\right) + \frac{\varepsilon}{2}\frac{d}{d\phi_\alpha}\left(f - \mu_{eq}c\right)\right\} + \frac{\varepsilon}{2}\left(\mu - \mu_{eq}\right)\frac{dc}{d\phi_\alpha}.$$

At equilibrium, $\mu = \mu_{eq}$, the second bracketed term has the form of a potential and when $\varepsilon f^*$ is of the same order or larger than $\gamma_{\alpha\beta}$, where $f^*$ is the energy scale, the potential scales with the length scale of the interface upon change in $\varepsilon$. In the sharp interface limit, the chemical potential at the leading order is constant across the interface, rendering the sharp interface limit for one dimensional evolution, as derived in Eqn.(6.5). We would derive a similar sharp interface limit with the grand potential functional in the absence of curvature. Note however, the effective potential in the case of the free energy functional scales with the interface width. This contribution leads to the modification of the surface energies and hence in the presence of curvature, the limits will differ. Additionally there is a deviation in the phase-field profile at leading order which is modified due to the presence of the grand potential excess as,

$$\frac{\partial\phi_\alpha}{\partial x} = -\frac{1}{\varepsilon}\sqrt{\frac{16}{\pi^2}\phi_\alpha\left(1 - \phi_\alpha\right) + \frac{\varepsilon}{T\gamma_{\alpha\beta}}\left(\left(f - \mu_{eq}c\right) - \left(f - \mu_{eq}c\right)_{bulk}\right)} \quad (8.4)$$

in contrast to $\dfrac{\partial\phi_\alpha}{\partial x} = -\dfrac{4}{\varepsilon\pi}\sqrt{\phi_\alpha\left(1 - \phi_\alpha\right)}$ for the case of the grand potential functional. This modifies, the effective kinetic coefficient applicable for both models. Therefore, to perform comparative simulations with the two models, the first challenge is to set the surface energies at leading order in the two models equivalent. The next, is to derive the desired interface widths to perform efficient simulations and finally to set the same kinetic coefficients for both models.

For the sake of discussion, consider the system Al-Cu created earlier in the chapter. While, in the case of the grand potential functional the interface width depends directly on the parameter $\varepsilon$, through the relation $\dfrac{\pi^2\varepsilon}{4} = 2.5\varepsilon$, and the surface energy is the same as the simulation parameter $T\gamma_{\alpha\beta}$; for the case of the free energy functional an expression for the interface width and surface energy at leading order can be derived as in Eqn.(6.3,6.4). Solving these equations simultaneously, we can derive the simulation parameters $\gamma_{\alpha\beta}$ and $\varepsilon$, to derive the surface energies $\tilde{\sigma}_{\alpha\beta}$ and $\tilde{\Lambda}_{\alpha\beta}$ for the given free energy functional. Whereas, this is possible for certain choices of $\varepsilon$, such as displayed in Fig. 8.16a, beyond a critical $\varepsilon$, there exists no unique solution, when the contribution to the potential from the grand potential excess, becomes dominant over that from the potential term $T\gamma_{\alpha\beta}\phi_\alpha\left(1-\phi_\alpha\right)$. The solution is then achieved for a range of $\gamma_{\alpha\beta}$ and $\varepsilon$, derived through the overlap of the isolevels of the $\tilde{\sigma}_{\alpha\beta}$ and the resulting $\tilde{\Lambda}_{\alpha\beta}$, which is fixed upon choosing the isolevel for the surface energy, Fig.8.16b. The computational efforts can be compared between the two models by estimating the interface widths used, when the same $\varepsilon$ is chosen for both the models. At the temperature T=0.988, Fig.8.17 displays the contours of a freely growing dendrite at a temperature of $T = 0.988$ ($T_m$=0.99) simulated using the grand potential formulation with $\varepsilon = 1688$. Corresponding to this $\varepsilon$, the interface width is $\tilde{\Lambda}_{\alpha\beta} = 430$ for the case of the free energy functional, Fig.8.16b. To have an interface resolution of 10 cells, would result in a grid resolution of $\Delta x = 43$ when simulations are performed using a regular grid. In contrast, for the case with a grand potential functional, we have used $\Delta x = 500$ for the simulation of the dendrite displayed in Fig.8.17, keeping the same interface resolution and conditions, which implies the computational effort increases $10^d$ times ($d$ denoting the dimension), when using the free energy functional. The situation is more favorable for the free energy functional at the higher undercooling at $T = 0.9843$ (smaller tip radii), as can be seen from Fig.8.16a, where the interface width for the case the free energy case results in 164 compared to a value of 281.5 for the grand potential functional at the same $\varepsilon = 112.5$. This is however, not the largest interface width that can be employed at this undercooling and much larger interface widths can be used for the case of the grand potential functional as was seen in Fig.8.2, while the interface widths that can be employed, when a free energy functional is used, gets limited to a smaller range. In summary, we conclude that at higher undercoolings(finer microstructures), the models based on the free energy functional and the grand potential functionals come closer. However at lower velocities or at lower undercoolings, the grid resolution

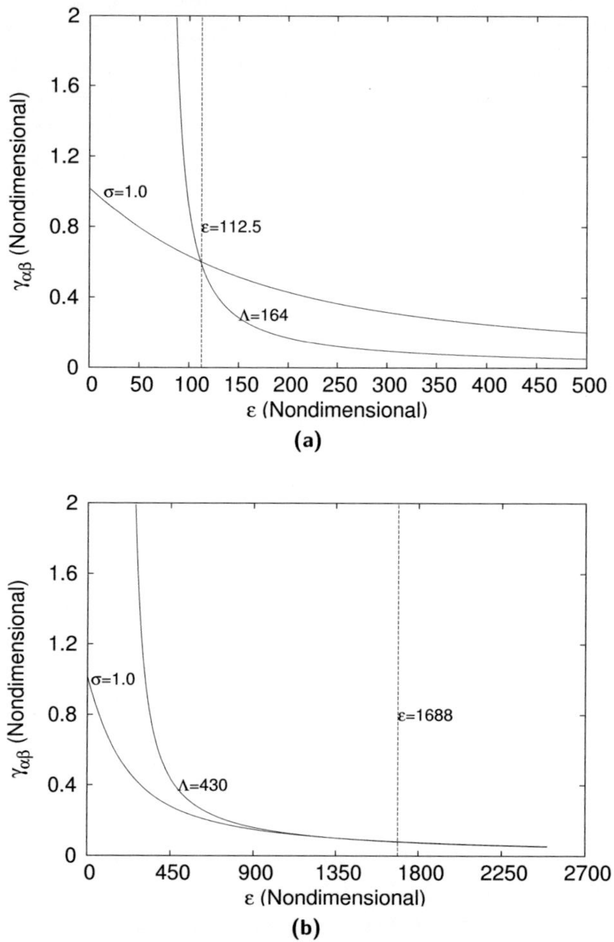


**Figure 8.16.:** (Please read $\sigma$ as $\tilde{\sigma}$ and $\Lambda$ as $\tilde{\Lambda}$). Defined contour level of the surface energy $\tilde{\sigma}$ and the interface width $\tilde{\Lambda}$ plotted as a function of the simulation parameters $\gamma_{\alpha\beta}$ and $\varepsilon$. In (a), the contours are calculated for the temperature $T = 0.9843$ while in (b) they are for a temperature of $T = 0.988$. The contours of the interface width are calculated from the defined level for the surface energy and the value of the $\varepsilon$ used in the simulation. All terms are dimensionless in the graphs.

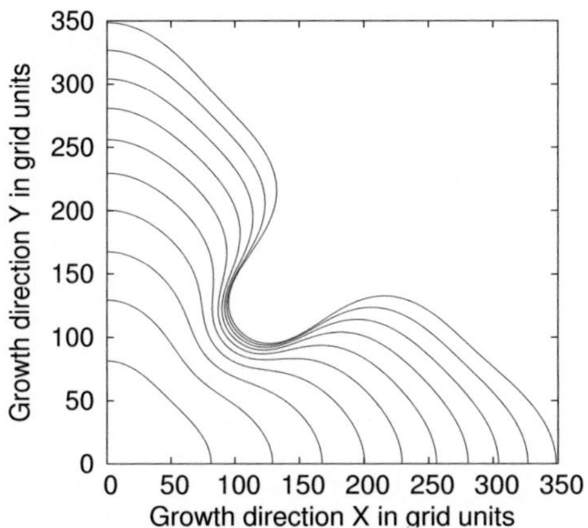


**Figure 8.17.:** Phase-field contours of a free growing equi-axed dendrite, with $\varepsilon = 1688$ at a temperature of T=0.988 ($T_m = 0.99$) with the grand potential formulation.

can be scaled up significantly using the grand potential functional which is however not a possibility using the free energy functional.

The comparison of the two models, is incomplete, without the discussion on the interface kinetics. As we have derived, larger interface widths can be employed for simulating the case of lower undercoolings with the grand potential functional. For such cases, the thin-interface limit is appropriate, and in this limit, the parameters can be chosen in such a manner, that interface kinetics vanishes which is relevant at lower undercoolings. To perform a similar asymptotics, for the free energy functional seems daunting and less useful because of the following reasoning. We recall that while performing the thin-interface asymptotics of the grand potential functional we have repeatedly used the anti-symmetric properties (odd functions) of the leading order phase-field profile $\phi_\alpha^0(x)$, which reduced the terms contributing to the first order correction to the chemical potential. However, with the free energy functional, the leading order solution is modified due to the grand potential excess and is derived using the Eqn.(8.4). Fig.8.18 compares the profiles in both cases, which shows slight asymmetry about the $\phi_\alpha = 0.5$ line, in the case of the free energy functional. The

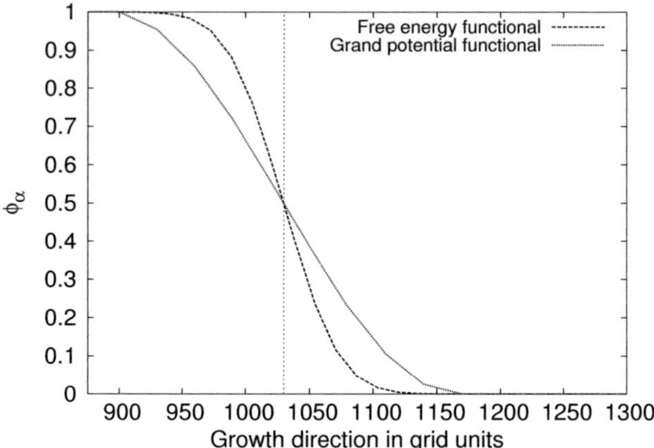


**Figure 8.18.:** Comparison of the leading order solution of the phase-field in the case of the grand potential functional and the free energy functional. The profiles have been superimposed and the vertical line representing the position of the binary interface is drawn for comparison.

magnitude of the asymmetry depends on the nature of the grand potential excess and scales with the interface widths. This asymmetry would rule out any general simplification of the solvability integrals appearing in the asymptotic analysis. Secondly, all solvability integrals depend on the leading order solution, which change with the interface widths, the temperature and the system one is simulating, thus certainly hindering, the universal applicability of the analysis. Lastly, due to the limitation of the interface widths, the thin interface limit is less applicable and performing simulations with vanishing interface kinetics is computationally expensive as the sharp interface and thin interface limits coincide for smaller interface widths.

To conclude the grand potential formulation offers significant flexibility in comparison to the free energy functionals because the range of applicability of the phase-field model is improved significantly without a corresponding increase in computational overhead and additionally, a thin-interface asymptotics with universal applicability can be performed.

## 8.4. Conclusions

We validate a newly derived model based on the grand potential functional. We test the asymptotics through comparison with analytical expressions for the growth of dendritic and eutectic growth morphologies and achieve reasonable agreement. With the derived model, simulation of large scale microstructures is definitely a possibility and the simulated dendritic structures look promising in this context.

One must note that the real gain with the modified model, is the range of undercoolings that can be spanned with numerical efficiency. For example, at higher undercoolings, the diffusion length is small which would need smaller grid resolution. This condition is apparently favorable when starting from a free energy functional, since the resultant small equilibrium interface thickness in the presence of high grand potential excess, can be resolved without enormous computational overhead. However, as we reduce the undercooling, the diffusion length becomes larger and one intends to use higher interface thicknesses to numerically resolve the microstructures. This flexibility is present, when one starts to derive from the grand potential model, while it becomes significantly expensive computationally, to resolve the same microstructure with the free energy functional(derived through the interpolation of the free energy densities of the phases as functions of the local concentration).

With regards to the asymptotics, the results obtained from the thin-interface asymptotics of model based on the grand potential functional, such as the antitrapping current and the expression for the interface coefficient are universally applicable for all undercoolings. Conversely, for the case of the model with the free energy functionals, the equilibrium properties related to the interface such as the surface energies and the interface thicknesses and the interface profiles at lowest order, are dependent on the undercooling. This implies that, expressions for the kinetic coefficients and the anti-trapping currents, if derived, would depend on the undercooling and hence, the solvability integrals would need to be re-calculated, given the undercooling one desires to simulate. In addition, the calibration of the simulation parameters to retrieve the desired surface tensions needs to be repeated for any change in the processing condition, which becomes highly cumbersome.

In summary, it is clear that, with the switch in the modeling ideology from the free energy functionals to a grand potential functional, one gains numerical flexibility and significant reduction in computational effort.

## 8.5. Outlook

The road is clear for the application of an efficient model for the study of various phenomena occurring during phase transformations in real alloys. For instance, in solidification the method can be applied for the investigation of dendritic and cellular growth, fragmentation, selection of dendritic and cell spacing and the cellular to dendritic transitions. Multiphase phase-transformations, involving eutectic, peritectic and monotectic solidification, in binary, ternary and higher systems are of interest. For instance, understanding pattern formation, in ternary eutectic systems under the influence of changes in the liquid compositions, imposed temperature gradients and velocities, during 3D microstructure evolution are some of the primary questions. In solid- state, the model can be tailored to treat stoichiometric compounds, such as the cementite phase in Fe-C alloy and studies of eutectoid coupled growth involving the growth of the cementite and ferrite from austenite, is a topic that can be attended. In addition, structural evolution during ripening of precipitates in different materials can also be treated fairly elegantly. In summary, the scope of applicability of the phase-field model has certainly increased with the suggested modifications. With additional developments in computational techniques such as 3D parallelization, adaptive mesh refinement techniques, large domain structures are certainly in the realm of the phase-field method.

While these are marked improvements, one must however, be aware of the limitations and assumptions that are present in the derivations and the limits of applicability of the model. For instance, the treatment of coupled fields such as temperature and concentration/chemical potential, require attention, as these involve the coupling of fields which evolve at different time and length scales. Similarly, including elastic, flow, magnetic and electric fields, although make the description more realistic, the cross-coupling between fields would need to be performed, keeping the required free boundary problem in mind. These are complex questions and certainly a challenge for the future.

# Appendix A

# Modeling phase diagrams

In the context of the phase-field method, there often arises the question as to how one can determine the model parameters given a free energy model. In this short discussion we look into this topic, considering ideal solution models for deriving simple isomorphous, eutectic and peritectic systems.

## A.1. Eutectics

Consider a binary eutectic system, with defined eutectic temperature $T_E$, volume fractions of the solid phases given by $\eta_\alpha$ and $\eta_\beta$ and the melting temperatures of the components A and B given by $T_A^\alpha$ and $T_B^\beta$ respectively where $\alpha$ is rich in A, and $\beta$ is the B-rich phase. We assume the non-dimensional free energies to be of the form:

$$f^\alpha(T,c) = cL_A^\alpha \frac{(T - T_A^\alpha)}{T_A^\alpha} + (1-c)L_B^\alpha \frac{(T - T_B^\alpha)}{T_B^\alpha} + T\left(c \ln c + (1-c)\ln(1-c)\right).$$

To determine the free energy parameters $L_A^\alpha$, $L_B^\alpha$, $L_A^\beta$, $L_B^\beta$ and the corresponding temperatures given by $T_A^\alpha$, $T_B^\alpha$, $T_A^\beta$, $T_B^\beta$, we write the equilibrium equations relevant at the eutectic temperature which are,

$$L_A^\alpha \frac{(T - T_A^\alpha)}{T_A^\alpha} + T \ln c^\alpha = T \ln c^l$$

$$L_B^\alpha \frac{(T - T_B^\alpha)}{T_B^\alpha} + T \ln(1 - c^\alpha) = T \ln(1 - c^l)$$

$$L_A^\beta \frac{\left(T - T_A^\beta\right)}{T_A^\beta} + T \ln c^\beta = T \ln c^l$$

$$L_B^\beta \frac{\left(T - T_B^\beta\right)}{T_B^\beta} + T \ln(1 - c^\beta) = T \ln(1 - c^l).$$

Since, we have eight unknowns and four equations, we require to make some assumptions. The values of $T_A^\alpha$ and $T_B^\beta$ are generally known, and hence the number of unknowns are reduced to six. One can make use of the assumption $L_A^\alpha = L_B^\alpha$ and $L_A^\beta = L_B^\beta$, which then reduces the required number of unknowns such that the equations can be solved to determine the free energy parameters. One can also derive the slopes of the liquidus and solidus of both phases and use them as the missing equations for the eight unknowns. The equations for the slopes of the liquidus of the $\alpha$ phase are as follows,

$$\frac{dT}{dc^l} = \frac{T^2 \left( \dfrac{c^\alpha}{c^l} - \dfrac{(1-c^\alpha)}{(1-c^l)} \right)}{(c^\alpha L_A^\alpha + (1-c^\alpha)L_B^\alpha)},$$

and similarly for the solidus, which can be written as,

$$\frac{dT}{dc^\alpha} = \frac{T^2 \left( \dfrac{c^l}{c^\alpha} - \dfrac{(1-c^l)}{(1-c^\alpha)} \right)}{(c^l L_A^\alpha + (1-c^l)L_B^\alpha)}.$$

Notice however, if we do not make any assumptions, we have the required number of equations for the number of unknowns. This implies, that we could end up with a solution for the terms $T_A^\alpha$ and $T_B^\beta$ which differ from reality. This is not problem though, since our data for the phase diagram is fitted accurately around the eutectic temperature, which suffices for simulations relevant for eutectics. The problem can however be tackled by considering temperature dependent terms $L_i^\alpha$, which however makes the system of equations not closed. The discussion on this is however, out of the scope of the present chapter.

## A.2. Peritectics

The discussion on the peritectics also stays very much the same as the eutectics. The only difference is, here only of the temperatures $T_\alpha^A$ is generally known. Depending, on the range of undercoolings one intends to simulate, this might be an important parameter to fit and the set of equilibrium equations along with the slopes of the corresponding solidus and liquidus can be solved around the peritectic temperature for all the unknowns.

## A.3. Isomorphous

If one is interested in fitting only two-phase equilibrium it can be done similarly by writing down the phase equilibrium equations around the temperature of interest. The slopes of the liquidus and solidus give the required equations for the relevant number of unknowns.

## A.4. Energy Scale

In the above discussion, only the data relevant to the phase-diagram are fitted. However, in order to derive the relevant coupling to the Gibbs-Thomson equation, it is important to set the energy scale of the system such that the right Gibbs-Thomson equations can be derived. This is easily done, by righting down the latent heat as $f^*T\left(\dfrac{\partial f^\alpha}{\partial T} - \dfrac{\partial f^l}{\partial T}\right)$ and then using the relation of the Gibbs-Thomson coefficient as $\Gamma = \dfrac{\tilde{\sigma}_{\alpha l}T}{L}$. This fixes a unique energy scale for the system which is no problem for the case of a two-phase equilibrium. However, if one treats three phase equilibria, this constraint implies, that the Gibbs-Thomson coefficients of the respective solid phases must obey a relationship among them given by,

$$\frac{\tilde{\sigma}_{\alpha l}}{\left(\dfrac{\partial f^\alpha}{\partial T} - \dfrac{\partial f^l}{\partial T}\right)\Gamma_\alpha} = \frac{\tilde{\sigma}_{\beta l}}{\left(\dfrac{\partial f^\beta}{\partial T} - \dfrac{\partial f^l}{\partial T}\right)\Gamma_\beta}.$$

While this might hold for many systems, for others the ideal solution model needs to be modified with temperature dependent terms $L_i^\alpha$.

Along with the energy scale the free energy of each phase becomes,

$$f^\alpha(c,T) = f^*\left(cL_A^\alpha\frac{(T - T_A^\alpha)}{T_A^\alpha} + \right.$$
$$\left.(1-c)L_B^\alpha\frac{(T - T_B^\alpha)}{T_B^\alpha} + T\left(c\ln c + (1-c)\ln(1-c)\right)\right).$$

# Appendix B

# Algorithm for calculation of critical nucleus

In the phase-field context, it sometimes becomes important to calculate the critical nucleus in a given concentration of the liquid at given temperature. For critical nuclei, beyond a size, the phase fractions go to 1 in the bulk, which can be determined from the Gibbs-Thomson condition $\Delta T = \Gamma \kappa$. While this is possible for the case of isotropic surface energies, the more general case of anisotropic surface energies is non-trivial. Also, the case of sub-critical nuclei cannot be treated with this method. To do this we derive a algorithm to compute the critical phase and concentration profiles, given a volume of nucleus. We utilize the concept of the "Volume Preserved Method", which is highlighted in Fig.B.1. The method can be

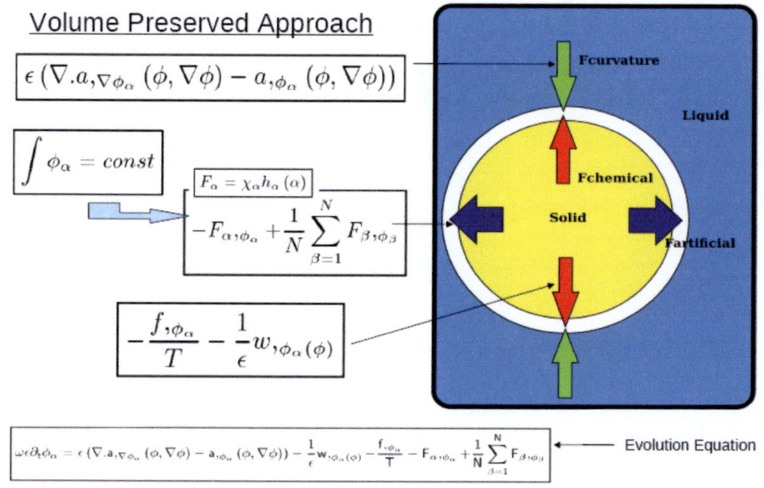


**Figure B.1.:** Procedure to calculate the critical nucleus by solving the Euler-Lagrange equations

very easily understood through this diagram, where the chemical and the capillary forces are shown as two opposing forces in evolution. The volume of the evolving particle is preserved by imposing the resultant of these two opposing forces with the direction reversed. This is realized through the construction of an artificial term $F_\alpha = \chi_\alpha$, where $\chi_\alpha$ is determined through the condition $\int \phi_\alpha = const$, $\alpha$ denoting the phase whose volume is preserved. For appropriate details one can refer to the article [80]. The

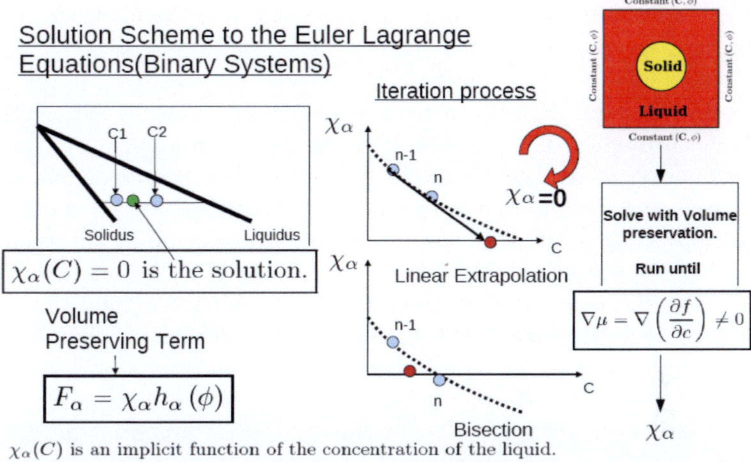



**Figure B.2.:** Procedure to calculate the critical nucleus by solving the Euler-Lagrange equations

method is used to derive the algorithm for determining the homogeneous critical nucleus potrayed in Fig.B.2. The aim lies in determining the critical concentration of the liquid given a volume of the phase $\alpha$ between the phase-stability lines of the phases $\alpha$ and the liquid phase. We start by the preserving the given volume of the solid while solving for the concentration profiles until the chemical potential gradient goes to zero with constant(Dirichlet) boundary conditions in all directions. At this time we determine the "Volume preservation force, $\chi_\alpha$". For the second iteration, we decide to modify the concentration in the liquid as $0.5(c_l^{eq}+c_l)$ or $0.5(c_s^{eq}+c_l)$ depending on whether the sign of $\chi_\alpha$ is positive or negative respectively. For the iterations henceforth, we may make the decision of the modification of concentrations based on two previous iterations. There exist two possibilities: 1)if the $\chi_\alpha^n$ and $\chi_\alpha^{n-1}$ are of the same sign, then the next estimate of the concentration $c_l$ is made by a linearly extrapolating the relation between $\chi_\alpha$ vs $c^l$ to zero. 2) If however, the signs of $\chi_\alpha$ from two previous iterations are opposite sign we perform a bisection and use this as the next estimate of the critical liquid concentration. This iteration process is continued until the term $\chi_\alpha$ goes to zero. The obtained solution satisfies the Euler-Lagrange equation representing $\dfrac{d\phi_\alpha}{dt} = 0$ and $\dfrac{dc}{dt} = 0$.

Figure B.3 displays results of nucleation during growth. Nucleation is imposed through stochastic noise in the bulk liquid in a Al-Cu alloy modeled using ideal free energies. The eutectic consists of two phases $\alpha-$Al rich and $\beta-$Cu rich. The barrier to nucleation for each phase are computed as a function of the critical composition in the liquid, by evaluating the grand potential excess with respect to the uniform initial liquid from the calculated critical solutions which are solutions to the Euler-Lagrange equations. We see that for the $\beta-$ phase the barrier to nucleation reduces as the concentration of Cu increases and vice-versa for the $\alpha-$ phase for which the barrier to nucleation increases with the increase in concentration of Cu. In the simulation domain, the dendrite growth occurs in a super-saturated liquid wih low concentration of Cu, and on imposition of noise the nucleation of the phase $\alpha-$ phase occurs in the far-field liquid. This corroborates well with the barrier to nucleation calculations, as the barrier to nucleation is lower for lower concentrations of Cu. As dendrite arms appear, the liquid entrapped in between the dendrite arms become eneriched in Cu, and the $\beta$ phase is seen to nucleate. This observation also qualitatively matches the inferences from the barrier to nucleation calculations.

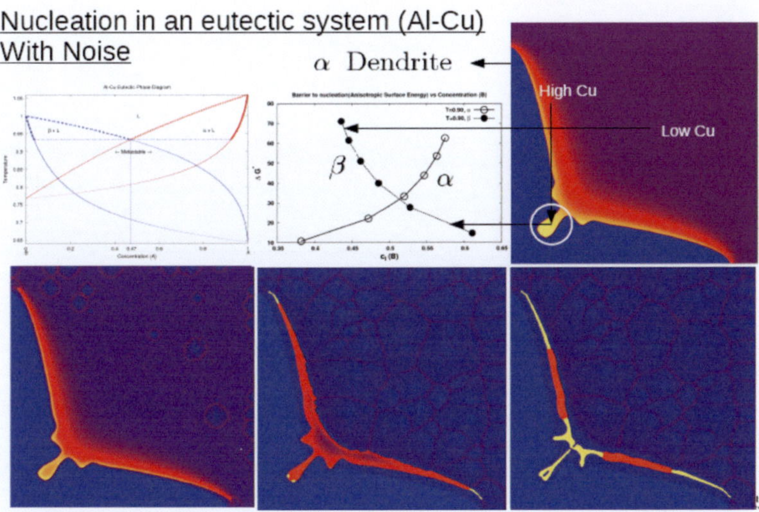


**Figure B.3.:** Nucleation of $\alpha$ nuclei in the bulk liquid and inter-dendritic eutectic phase in between the dendritic arms

# Appendix C

# Correction to interpolation polynomials

Phase field modeling has spread to the solution of a variety of problems. The solution methodology is lucrative because it does not need the tracking of evolving interfaces. The basis of the phase field theory lies in treating each evolving phase with an *order parameter* and describing its evolution on the basis of minimization of energy or the maximization of entropy of the whole system. The evolution of the order parameters are coupled with the evolution of other fields which are for example the concentration and the temperature. One of the components required for the description of the energy or the entropy density, is that either of these quantities be decribed in the whole phase space. This is done by the use of *interpolation functions* which interpolate between the values in the bulk phases. Thus for e.g we have

$$F = F_\alpha h_\alpha\left(\phi\right) + F_\beta h_\beta\left(\phi\right) + F_\gamma h_\gamma\left(\phi\right)\ldots$$

where $F_\alpha$, $F_\beta$, $F_\gamma$ are the bulk properties of the $\alpha$, $\beta$ and the $\gamma$ phases respectively. The functions $h_\alpha$, $h_\beta$ and $h_\gamma$ are the functions interpolating the bulk properties of the respective phases. The vector $\phi$ belongs to the N-1 dimensional space

$$\sum_i^N \phi_i = 1 \tag{C.1}$$

$\phi = (\phi_\alpha, \phi_\beta, \phi_\gamma \ldots)$, N being the number of phases in the system. This results from treating the order parameters as volume fractions of the phases in the system. It can be seen in the above description that the interpolation functions are averaging functions for the bulk properties of the system, with weights depending on where one is in the phase space. Thus, it is natural to have the sum of the weights or the sum of the interpolation functions to add up to 1. One of the topics extensively studied using the phase field methodolgy is *solidification* of solid phase from a liquid phase, and two interpolation functions have been widely used for this purpose,

$$h_\alpha\left(\phi\right) = \phi_\alpha^2\left(3 - 2\phi_\alpha\right) \tag{C.2}$$

$$h_\alpha\left(\phi\right) = \phi_\alpha^3\left(10 - 15\phi_\alpha + 6\phi_\alpha^2\right) \tag{C.3}$$

The above functions satisfy the summation property for two phases, but fail to do so as the number of phases increases in number. So there is a need for finding correct interpolation functions when modeling more than two phases. In the following paper we suggest ways of correcting such interpolation functions for use in the case of more than two phases.

## C.1. The correction to the third order polynomial

The cubic third order polynomial for two phases $(\phi_\alpha, \phi_\beta)$ can be created very easily in the following manner.

$$(\phi_\alpha + \phi_\beta)^3 = \phi_\alpha^3 + \phi_\beta^3 + 3\phi_\alpha^2\phi_\beta + 3\phi_\beta^2\phi_\alpha$$

We now separate the R.H.S into two symmetric parts and call each of them an interpolation function. There are two possible ways to choose the symmetric functions, we choose the one which gives a function that has the lowest order of $\phi_\nu$ in $h_\nu(\phi)$ greater than equal to 2, i.e

$$h_\alpha(\phi) = \phi_\alpha^3 + 3\phi_\alpha^2\phi_\beta$$
$$h_\beta(\phi) = \phi_\beta^3 + 3\phi_\beta^2\phi_\alpha$$

. The choice is due to the following reason. We want to have

$$\left(\frac{\partial h_\nu(\phi)}{\partial \phi_\nu}\right)_{\phi_\alpha+\phi_\beta=1} = \frac{\partial h_\nu(\phi)}{\partial \phi_\nu} - \frac{1}{N}\sum_k^N \frac{\partial h_\nu(\phi)}{\partial \phi_k}$$

$=0$, at $\phi_\nu = 0, 1$. The L.H.S of the above equation is the deriviative with respect to $\phi_\nu$ with the constraint eqn C.1. Now using the constraint $\phi_\alpha + \phi_\beta = 1$, we have the interpolation functions which were described before eqn C.2. We also have that $h_\nu(\phi) = 0, 1$ respectively for any point in the phase space with $\phi_\nu = 0, 1$ respectively. The above deriviation shows that the summation over all the cubic interpolation functions of the above form, at any point in the phase space equals 1 only in the case of two phases. We propose a correction to the interpolation function by doing the same procedure for more than two phases. We first do it for three phases and extend it to more than three phases.

$$(\phi_\alpha + \phi_\beta + \phi_\gamma)^3 = \phi_\alpha^3 + \phi_\beta^3 + \phi_\gamma^3 + 3\phi_\alpha^2 (\phi_\beta + \phi_\gamma) +$$
$$3\phi_\beta^2 (\phi_\alpha + \phi_\gamma) + 3\phi_\gamma^2 (\phi_\alpha + \phi_\beta) + 6\phi_\alpha\phi_\beta\phi_\gamma$$

Again we try to find three symmetric terms in the above sum, at all points trying to have $h_\nu(\phi)$ with the highest power of $\phi_\nu$. We find that in this case the lowest order possible has to be linear in $\phi_\nu$ for $h_\nu(\phi)$. So using the sum constraint for the phases, the interpolation functions can be written in the following form,

$$h_\alpha(\phi) = \phi_\alpha^2 (3 - 2\phi_\alpha) + 2\phi_\alpha\phi_\beta\phi_\gamma$$
$$h_\beta(\phi) = \phi_\beta^2 (3 - 2\phi_\beta) + 2\phi_\alpha\phi_\beta\phi_\gamma$$
$$h_\gamma(\phi) = \phi_\gamma^2 (3 - 2\phi_\gamma) + 2\phi_\alpha\phi_\beta\phi_\gamma$$

The above procedure can be repeated for N phases and we would have the following function.

$$h_\nu(\phi) = \phi_\nu^2 (3 - 2\phi_\nu) + 2\phi_\nu \left( \sum_{\beta < \gamma (\beta, \gamma \neq \nu)} \phi_\beta\phi_\gamma \right)$$

## C.2. The correction to the fifth order polynomial

The correction to fifth order polynomial follows in the same manner as in the case of the cubic polynomial. We try to derive a polynomial using the above procedure for three phases,

$$(\phi_\alpha + \phi_\beta + \phi_\gamma)^5 = \phi_\alpha^5 + \phi_\beta^5 + \phi_\gamma^5 + 5\phi_\alpha^4 (1 - \phi_\alpha) + 5\phi_\beta^4 (1 - \phi_\beta) +$$
$$5\phi_\gamma^4 (1 - \phi_\gamma) + 10\phi_\alpha^3 (1 - \phi_\alpha)^2 + 10\phi_\beta^3 (1 - \phi_\beta)^2 + 10\phi_\gamma^3 (1 - \phi_\gamma)^2$$

$$+ 30\phi_\alpha\phi_\beta^2\phi_\gamma^2 + 30\phi_\beta\phi_\alpha^2\phi_\gamma^2 + 30\phi_\gamma\phi_\alpha^2\phi_\beta^2$$

Now we divide the sum into three parts in such a way that the least power of $\phi_\nu$ in $h_\nu(\boldsymbol{\phi})$ is 2. This results in the following interpolation functions,

$$h_\alpha(\boldsymbol{\phi}) = \phi_\alpha^5 + 5\phi_\alpha^4(1 - \phi_\alpha) + 10\phi_\alpha^3(1 - \phi_\alpha)^2 + 15\phi_\alpha^2\phi_\gamma\phi_\beta(\phi_\beta + \phi_\gamma)$$

$$h_\beta(\boldsymbol{\phi}) = \phi_\beta^5 + 5\phi_\beta^4(1 - \phi_\beta) + 10\phi_\beta^3(1 - \phi_\beta)^2 + 15\phi_\beta^2\phi_\gamma\phi_\alpha(\phi_\alpha + \phi_\gamma)$$

$$h_\gamma(\boldsymbol{\phi}) = \phi_\gamma^5 + 5\phi_\gamma^4(1 - \phi_\gamma) + 10\phi_\gamma^3(1 - \phi_\gamma)^2 + 15\phi_\gamma^2\phi_\alpha\phi_\beta(\phi_\alpha + \phi_\beta)$$

We simplify the terms in the following manner

$$\phi_\alpha\phi_\beta = \frac{\left((\phi_\alpha + \phi_\beta)^2 - (\phi_\alpha - \phi_\beta)^2\right)}{4} = \frac{\left((1 - \phi_\gamma)^2 - (\phi_\alpha - \phi_\beta)^2\right)}{4}$$

$$\phi_\alpha\phi_\gamma = \frac{\left((\phi_\gamma + \phi_\alpha)^2 - (\phi_\gamma - \phi_\alpha)^2\right)}{4} = \frac{\left((1 - \phi_\beta)^2 - (\phi_\gamma - \phi_\alpha)^2\right)}{4}$$

$$\phi_\gamma\phi_\beta = \frac{\left((\phi_\gamma + \phi_\beta)^2 - (\phi_\gamma - \phi_\beta)^2\right)}{4} = \frac{\left((1 - \phi_\alpha)^2 - (\phi_\gamma - \phi_\beta)^2\right)}{4}$$

Substituting in the above interpolation functions we have,

$$h_\alpha(\boldsymbol{\phi}) = \phi_\alpha^5 + 5\phi_\alpha^4(1 - \phi_\alpha) + 10\phi_\alpha^3(1 - \phi_\alpha)^2 + \frac{15}{4}\phi_\alpha^2(1 - \phi_\alpha)^3 -$$
$$\frac{15}{4}\phi_\alpha^2(1 - \phi_\alpha)(\phi_\gamma - \phi_\beta)^2$$

$$h_\beta(\boldsymbol{\phi}) = \phi_\beta^5 + 5\phi_\beta^4(1 - \phi_\beta) + 10\phi_\beta^3(1 - \phi_\beta)^2 + \frac{15}{4}\phi_\beta^2(1 - \phi_\beta)^3 -$$
$$\frac{15}{4}\phi_\beta^2(1 - \phi_\beta)(\phi_\alpha - \phi_\gamma)^2$$

$$h_\gamma(\boldsymbol{\phi}) = \phi_\gamma^5 + 5\phi_\gamma^4(1 - \phi_\gamma) + 10\phi_\gamma^3(1 - \phi_\gamma)^2 + \frac{15}{4}\phi_\gamma^2(1 - \phi_\gamma)^3 -$$
$$\frac{15}{4}\phi_\gamma^2(1 - \phi_\gamma)(\phi_\alpha - \phi_\beta)^2$$

The above are the interpolation functions suggested by Plapp et al. [32]. Extending the analysis for four phase space$(\phi_\alpha, \phi_\beta, \phi_\gamma, \phi_\delta)$ and repeating the same procedure as above we get the following form for the interpolation functions,

$$
\begin{aligned}
h_\alpha(\boldsymbol{\phi}) =\ & \phi_\alpha^5 + 5\phi_\alpha^4(1-\phi_\alpha) + 10\phi_\alpha^3(1-\phi_\alpha)^2 + \\
& 15\phi_\alpha^2\left(\phi_\gamma\phi_\beta(\phi_\beta+\phi_\gamma) + \phi_\gamma\phi_\delta(\phi_\gamma+\phi_\delta) + \phi_\beta\phi_\gamma(\phi_\beta+\phi_\gamma)\right) + \\
& 60\phi_\alpha^2\phi_\beta\phi_\gamma\phi_\delta \\
h_\beta(\boldsymbol{\phi}) =\ & \phi_\beta^5 + 5\phi_\beta^4(1-\phi_\beta) + 10\phi_\beta^3(1-\phi_\beta)^2 + \\
& 15\phi_\beta^2\left(\phi_\alpha\phi_\gamma(\phi_\alpha+\phi_\gamma) + \phi_\gamma\phi_\delta(\phi_\gamma+\phi_\delta) + \phi_\alpha\phi_\delta(\phi_\alpha+\phi_\delta)\right) + \\
& 60\phi_\beta^2\phi_\alpha\phi_\gamma\phi_\delta \\
h_\gamma(\boldsymbol{\phi}) =\ & \phi_\gamma^5 + 5\phi_\gamma^4(1-\phi_\gamma) + 10\phi_\gamma^3(1-\phi_\gamma)^2 + \\
& 15\phi_\gamma^2\left(\phi_\alpha\phi_\beta(\phi_\alpha+\phi_\beta) + \phi_\alpha\phi_\delta(\phi_\alpha+\phi_\delta) + \phi_\beta\phi_\delta(\phi_\beta+\phi_\delta)\right) + \\
& 60\phi_\gamma^2\phi_\alpha\phi_\beta\phi_\delta \\
h_\delta(\boldsymbol{\phi}) =\ & \phi_\delta^5 + 5\phi_\delta^4(1-\phi_\delta) + 10\phi_\delta^3(1-\phi_\delta)^2 + \\
& 15\phi_\delta^2\left(\phi_\alpha\phi_\beta(\phi_\alpha+\phi_\beta) + \phi_\alpha\phi_\gamma(\phi_\alpha+\phi_\gamma) + \phi_\beta\phi_\gamma(\phi_\beta+\phi_\gamma)\right) + \\
& 60\phi_\delta^2\phi_\alpha\phi_\beta\phi_\delta
\end{aligned}
$$

The above expressions can be simplified in the following manner,

$$
\begin{aligned}
h_\alpha(\boldsymbol{\phi}) =\ & \phi_\alpha^5 + 5\phi_\alpha^4(1-\phi_\alpha) + 10\phi_\alpha^3(1-\phi_\alpha)^2 + \\
& 15\phi_\alpha^2(1-\phi_\alpha)(\phi_\gamma\phi_\beta + \phi_\gamma\phi_\delta + \phi_\beta\phi_\gamma) + 15\phi_\alpha^2\phi_\beta\phi_\gamma\phi_\delta \\
h_\beta(\boldsymbol{\phi}) =\ & \phi_\beta^5 + 5\phi_\beta^4(1-\phi_\beta) + 10\phi_\beta^3(1-\phi_\beta)^2 + \\
& 15\phi_\beta^2(1-\phi_\beta)(\phi_\gamma\phi_\alpha + \phi_\gamma\phi_\delta + \phi_\alpha\phi_\delta) + 15\phi_\beta^2\phi_\alpha\phi_\gamma\phi_\delta \\
h_\gamma(\boldsymbol{\phi}) =\ & \phi_\gamma^5 + 5\phi_\gamma^4(1-\phi_\gamma) + 10\phi_\gamma^3(1-\phi_\gamma)^2 + \\
& 15\phi_\gamma^2(1-\phi_\gamma)(\phi_\alpha\phi_\beta + \phi_\beta\phi_\delta + \phi_\alpha\phi_\delta) + 15\phi_\gamma^2\phi_\alpha\phi_\beta\phi_\delta \\
h_\delta(\boldsymbol{\phi}) =\ & \phi_\delta^5 + 5\phi_\delta^4(1-\phi_\delta) + 10\phi_\delta^3(1-\phi_\delta)^2 + \\
& 15\phi_\delta^2(1-\phi_\delta)(\phi_\alpha\phi_\beta + \phi_\alpha\phi_\gamma + \phi_\gamma\phi_\beta) + 15\phi_\delta^2\phi_\alpha\phi_\beta\phi_\gamma
\end{aligned}
$$

Writing the following terms in a analogous manner we get,

$$
(\phi_\gamma\phi_\beta + \phi_\gamma\phi_\delta + \phi_\beta\phi_\gamma) =
$$

$$\frac{\left(2\left(\phi_\beta+\phi_\gamma+\phi_\delta\right)^2-\left(\phi_\beta-\phi_\gamma\right)^2-\left(\phi_\gamma-\phi_\delta\right)^2-\left(\phi_\beta-\phi_\delta\right)^2\right)}{6}$$

$$=\frac{\left(2\left(1-\phi_\alpha\right)^2-\left(\phi_\beta-\phi_\gamma\right)^2-\left(\phi_\gamma-\phi_\delta\right)^2-\left(\phi_\beta-\phi_\delta\right)^2\right)}{6}$$

$$\left(\phi_\gamma\phi_\alpha+\phi_\gamma\phi_\delta+\phi_\alpha\phi_\delta\right)=$$

$$\frac{\left(2\left(\phi_\alpha+\phi_\gamma+\phi_\delta\right)^2-\left(\phi_\alpha-\phi_\gamma\right)^2-\left(\phi_\gamma-\phi_\delta\right)^2-\left(\phi_\alpha-\phi_\delta\right)^2\right)}{6}$$

$$=\frac{\left(2\left(1-\phi_\beta\right)^2-\left(\phi_\alpha-\phi_\gamma\right)^2-\left(\phi_\gamma-\phi_\delta\right)^2-\left(\phi_\beta-\phi_\delta\right)^2\right)}{6}$$

$$\left(\phi_\alpha\phi_\beta+\phi_\beta\phi_\delta+\phi_\alpha\phi_\delta\right)=$$

$$\frac{\left(2\left(\phi_\alpha+\phi_\beta+\phi_\delta\right)^2-\left(\phi_\alpha-\phi_\beta\right)^2-\left(\phi_\beta-\phi_\delta\right)^2-\left(\phi_\alpha-\phi_\delta\right)^2\right)}{6}$$

$$=\frac{\left(2\left(1-\phi_\gamma\right)^2-\left(\phi_\alpha-\phi_\beta\right)^2-\left(\phi_\beta-\phi_\delta\right)^2-\left(\phi_\gamma-\phi_\delta\right)^2\right)}{6}$$

$$\left(\phi_\alpha\phi_\beta+\phi_\alpha\phi_\gamma+\phi_\gamma\phi_\beta\right)=$$

$$\frac{\left(2\left(\phi_\alpha+\phi_\beta+\phi_\gamma\right)^2-\left(\phi_\alpha-\phi_\beta\right)^2-\left(\phi_\beta-\phi_\gamma\right)^2-\left(\phi_\alpha-\phi_\gamma\right)^2\right)}{6}$$

$$=\frac{\left(2\left(1-\phi_\delta\right)^2-\left(\phi_\alpha-\phi_\beta\right)^2-\left(\phi_\beta-\phi_\gamma\right)^2-\left(\phi_\alpha-\phi_\gamma\right)^2\right)}{6}$$

Substituting in the interpolation functions, the functions can be written as the following,

$$h_\nu\left(\phi\right)=\phi_\nu^5+5\phi_\nu^4\left(1-\phi_\nu\right)+10\phi_\nu^3\left(1-\phi_\nu\right)^2+5\phi_\nu^2\left(1-\phi_\nu\right)^3-$$

$$\frac{5}{2}\phi_\nu^2\left(1-\phi_\nu\right)\left(\sum_{\beta<\gamma,\beta,\gamma\neq\nu}\left(\phi_\beta-\phi_\gamma\right)^2\right)$$

$$+15\phi_\nu^2\phi_\beta\phi_\gamma\phi_\delta\quad\forall\nu,\left(\beta,\gamma,\delta\neq\nu\right)$$

The process when done for five phases gives the following form of the interpolation function,

$$h_\nu\left(\phi\right) = \phi_\nu^5 + 5\phi_\nu^4\left(1 - \phi_\nu\right) + 10\phi_\nu^3\left(1 - \phi_\nu\right)^2 + \frac{45}{8}\phi_\nu^2\left(1 - \phi_\nu\right)^3 -$$

$$\frac{15}{8}\phi_\nu^2\left(1 - \phi_\nu\right)\left(\sum_{\beta<\gamma,\beta,\gamma\neq\nu}\left(\phi_\beta - \phi_\gamma\right)^2\right)$$

$$+ 15\phi_\nu^2\left(\sum_{\beta<\gamma<\delta,\beta,\delta,\gamma\neq\nu}\phi_\beta\phi_\gamma\phi_\delta\right) + 24\phi_\nu\phi_\alpha\phi_\beta\phi_\gamma\phi_\delta$$

$$\forall\nu,\left(\alpha,\beta,\gamma,\delta\neq\nu\right)$$

The process can now be generalized for N phases, and the interpolation function looks like,

$$h_\nu\left(\phi\right) = \phi_\nu^5\left(6 - \frac{15\left(N - 2\right)}{2\left(N - 1\right)}\right) + \phi_\nu^4\left(-15 + \frac{45\left(N - 2\right)}{2\left(N - 1\right)}\right) +$$

$$\phi_\nu^3\left(10 + \frac{15}{2\left(N - 1\right)}\left(\sum_{\beta<\gamma(\beta,\gamma\neq\nu)}\left(\phi_\beta - \phi_\gamma\right)^2\right) - \frac{45}{2}\frac{N - 2}{N - 1}\right) +$$

$$\phi_\nu^2\left(\frac{15\left(N - 2\right)}{2\left(N - 1\right)} - \frac{15}{2\left(N - 1\right)}\left(\sum_{\beta<\gamma(\beta,\gamma\neq\nu)}\left(\phi_\beta - \phi_\gamma\right)^2\right) +\right.$$

$$\left.15\left(\sum_{\beta<\gamma<\delta(\delta,\beta,\gamma\neq\nu)}\phi_\delta\phi_\beta\phi_\gamma\right)\right) + 24\phi_\nu\left(\sum_{\beta<\gamma<\delta<\alpha(\alpha,\delta,\beta,\gamma\neq\nu)}\phi_\delta\phi_\beta\phi_\gamma\phi_\alpha\right)$$

# Appendix D

# Choosing simulation parameters

Performing quantitative simulations requires one to have to control over the simulation parameters and the resulting physical quantities they represent. To achieve this a throrough knowledge of the phase-field model and asymptotics is essential. In this chapter, we look into the selection of simulation parameters $\gamma_{\alpha\beta}$ and $\varepsilon$ when setting the surface energies $\tilde{\sigma}_{\alpha\beta}$ and interface width $\tilde{\Lambda}_{\alpha\beta}$ for the case when the phase-field model with a free-energy functional is used. To start with one can derive the quantities $\tilde{\sigma}_{\alpha\beta}$ and $\tilde{\Lambda}_{\alpha\beta}$ given the free energy landscape in the following manner:

$$\tilde{\sigma}_{\alpha\beta} = 2\gamma_{\alpha\beta}T \int_0^1 \sqrt{\frac{16}{\pi^2}\phi_\alpha\left(1-\phi_\alpha\right) + \frac{\varepsilon}{\gamma_{\alpha\beta}T}\Delta\Psi\left(T,\boldsymbol{c},\boldsymbol{\phi}\right)}d\phi_\alpha \qquad \text{(D.1)}$$

$$\tilde{\Lambda}_{\alpha\beta} = \varepsilon \int_0^1 \frac{d\phi_\alpha}{\sqrt{\frac{16}{\pi^2}\phi_\alpha\left(1-\phi_\alpha\right) + \frac{\varepsilon}{\gamma_{\alpha\beta}T}\Delta\Psi\left(T,\boldsymbol{c},\boldsymbol{\phi}\right)}}, \qquad \text{(D.2)}$$

where $\Delta\Psi\left(T,\boldsymbol{c},\boldsymbol{\phi}\right)$ is the grand potential excess across the interface between two bulk phases at equilibrium given as, $\left(f\left(T,\boldsymbol{c},\boldsymbol{\phi}\right) - \sum_{i=1}^{K-1}\mu_i c_i\right) - \left(f\left(T,\boldsymbol{c},\boldsymbol{\phi}\right) - \sum_{i=1}^{K-1}\mu_i c_i\right)_{\phi_\alpha=0}$. To determine the simulation parameters $\gamma_{\alpha\beta}$ the two Eqns. D.1 and D.2, need to be solved simultaneosly. While this is possible for the case where when the grand potential excess is small D.1a, in other cases there exist infinite solutions for $\gamma_{\alpha\beta}$ and $\varepsilon$ along the intersecting isolines for $\tilde{\sigma}_{\alpha\beta} = K1$ and $\tilde{\Lambda} = K2$, where $K1$, $K2$ are required surface energies and the interface widths desired in the simulations D.1b. One must note, that in this case, once, the surface tension is fixed, there exists only one interface width $\tilde{\Lambda}_{\alpha\beta}$ and hence is no longer a degree of freedom. In the case of of multi-phases, the surface energies and the interface widths of all the interfaces cannot be fixed independently with just one parameter $\varepsilon$. In such cases, it makes sense to resolve, the smallest interface with the required number of grid points. With this value of $\varepsilon$, the parameter $\gamma_{\alpha\beta}$ of the other interfaces can be derived using the expression of the surface energy D.1.

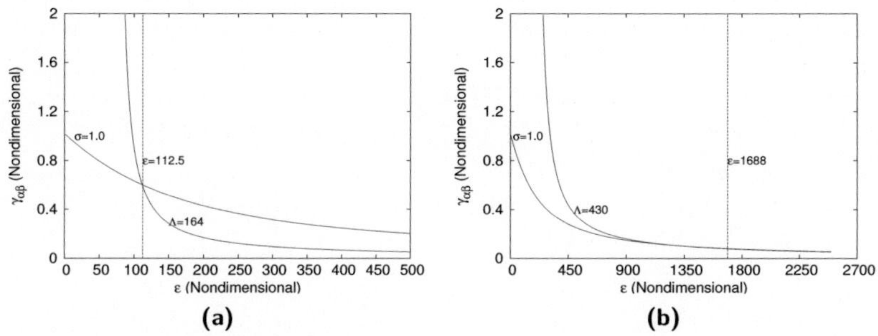


**Figure D.1.:** Two possibilities that might arise when treating real systems. In (a), the simulation parameters $\gamma_{\alpha\beta}$ and $\varepsilon$ can be fixed uniquely, the solution given by the intersection of the isolines $\tilde{\sigma}_{\alpha\beta} = const$ and $\tilde{\Lambda} = const$. When the magnitude of the grand potential excess goes higher, there no longer exists a unique solution as in (b) and a range of solutions exists which is given by the overlap of the required isolines of $\tilde{\sigma}_{\alpha\beta}$ and $\tilde{\Lambda}_{\alpha\beta}$.

# D.1. Example

Consider a binary eutectic system of three phases modeled with idealized free energies of the form,

$$f^{\alpha}(T,c) =$$

$$cL^{\alpha}\frac{(T - T_A^{\alpha})}{T_A^{\alpha}} + (1 - c)L_B^{\alpha}\frac{(T - T^B)}{T_{\alpha}^B} + T(c\ln c + (1 - c)\ln(1 - c))$$

with the parameters for the free energy $L_i^{\alpha}$ given as, $L_A^{\alpha} = L_B^{\alpha} = L_A^{\beta} = L_B^{\beta} = 5.0$, $T_i^{\alpha}$ given as, $T_A^{\alpha} = T_B^{\beta} = 1.0$ and $T_B^{\alpha} = T_A^{\beta} = 0.72318$. With these values, the simulation parameters of the solid-solid interface are derived as $\gamma_{\alpha\beta} = 0.285224$ and $\varepsilon = 6.3252$, to retrieve a surface energy $\tilde{\sigma}_{\alpha\beta} = 1.0$ and $\tilde{\Lambda}_{\alpha\beta} = 10.0$. For the other interfaces, we derive the surface energies $\tilde{\sigma}_{\alpha l} = \tilde{\sigma}_{\beta l} = 1.0$ using the value for $\varepsilon = 6.3252$ and calculating, $\gamma_{\alpha l} = \gamma_{\beta l} = 0.7254328$. The system of three phases are set at the critical undercooling given by $\Delta T = \Gamma_{\alpha l}\kappa = \Gamma_{\beta l}\kappa$, where $\kappa = \dfrac{2\sin\theta}{\lambda}$ and $\lambda$ is half the box-width that one is simulating. The angles at equilibrium are measured by fitting circles as shown below, and were calculated as 120.33,

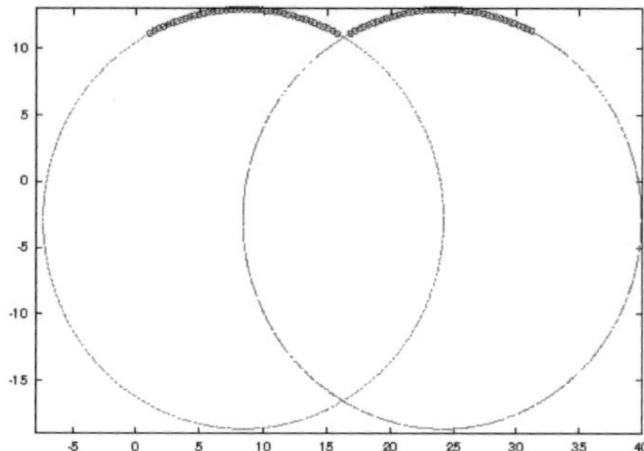

**Figure D.2.:** Circle fit of the both solid-liquid interfaces at equilibrium, when the system is set at the critical undercooling

which compares well to the theoretical prediction given by the Young's equilibrium condition.

# Appendix E

# Discretization of the antitrapping current

The anti-trapping current is derived to be of the form,

$$
\dot{\jmath}_{at} = -\frac{\pi\varepsilon}{4}\frac{g_\alpha\left(\phi_\alpha\right)\left(1-h_\alpha\left(\phi_\alpha\right)\right)}{\sqrt{\phi_\alpha\left(1-\phi_\alpha\right)}}\left(c^\beta\left(\mu,T\right)-c^\alpha\left(\mu,T\right)\right)\frac{\partial\phi_\alpha}{\partial t}\frac{\nabla\phi_\alpha}{|\nabla\phi_\alpha|}. \tag{E.1}
$$

where $g_\alpha\left(\phi_\alpha\right)$ and $h_\alpha\left(\phi_\alpha\right)$ are interpolation functions interpolating the diffisivities and the grand potentials respectively. The anti-trapping current enters the diffusion equation in the following form,

$$
\frac{\partial c}{\partial t} = \nabla\cdot\left(M\left(T,c,\phi\right)\nabla\mu - \jmath_{at}\right). \tag{E.2}
$$

The aim with the dicretization is to have all quantities computed at grid-point $i,j,k$ with second order accuracy in space. For this, gradients of the chemical potential are computed with second order accuracy at off-grid positions shifted by half the cell-size in each direction. Similarly, the values of $M\left(T,c,\phi\right)$ are determined at the off-grid positions by averaging the values of the neighboring cells in each direction. A description of this is given below,

$$
M\left(T,c,\phi\right)\nabla_x\mu^{i,j,k} = \frac{1}{2}\Big(M\left(T^{i,j,k},c^{i,j,k},\phi^{i,j,k}\right) +
$$
$$
M\left(T^{i+1,j,k},c^{i+1,j,k},\phi^{i+1,j,k}\right)\Big)\frac{\left(\mu^{i+1,j,k}-\mu^{i,j,k}\right)}{\Delta x}
$$
$$
M\left(T,c,\phi\right)\nabla_y\mu^{i,j,k} = \frac{1}{2}\Big(M\left(,T^{i,j,k},c^{i,j,k},\phi^{i,j,k}\right) +
$$
$$
M\left(T^{i,j+1,k},c^{i,j+1,k},\phi^{i,j+1,k}\right)\Big)\frac{\left(\mu^{i,j+1,k}-\mu^{i,j,k}\right)}{\Delta y}
$$
$$
M\left(T,c,\phi\right)\nabla_z\mu^{i,j,k} = \frac{1}{2}\Big(M\left(,T^{i,j,k},c^{i,j,k},\phi^{i,j,k}\right) +
$$
$$
M\left(T^{i,j,k+1},c^{i,j,k+1},\phi^{i,j,k+1}\right)\Big)\frac{\left(\mu^{i,j,k+1}-\mu^{i,j,k}\right)}{\Delta z}.
$$

With this the gradients in each direction are second-order accurate at offset grid positions, given by $\left(i+1/2,j,k\right),\left(i,j+1/2,k\right),\left(i,j,k+1/2\right)$ in directions $x,y,z$ respectively.

Similarly, we require the anti-trapping current at the off-grid positions, hence we perform the following discretization,

$$
\left(j_{at}^x\right)_{i+\frac{1}{2},j,k} = \left[-\frac{\pi\varepsilon}{4}\frac{g_\alpha\left(\phi_\alpha\right)\left(1-h_\alpha\left(\phi_\alpha\right)\right)}{\sqrt{\phi_\alpha\left(1-\phi_\alpha\right)}}\left(c^\beta\left(\mu,T\right)-c^\alpha\left(\mu,T\right)\right)\frac{\partial\phi_\alpha}{\partial t}\right]_{i+\frac{1}{2},j,k}
$$

$$
\times\frac{\left(\nabla_x\phi_\alpha\right)_{i+\frac{1}{2},j,k}}{\left|\nabla\phi_\alpha\right|_{i+1/2,j,k}}
$$

$$
\left(j_{at}^y\right)_{i,j+\frac{1}{2},k} = \left[-\frac{\pi\varepsilon}{4}\frac{g_\alpha\left(\phi_\alpha\right)\left(1-h_\alpha\left(\phi_\alpha\right)\right)}{\sqrt{\phi_\alpha\left(1-\phi_\alpha\right)}}\left(c^\beta\left(\mu,T\right)-c^\alpha\left(\mu,T\right)\right)\frac{\partial\phi_\alpha}{\partial t}\right]_{i,j+\frac{1}{2},k}
$$

$$
\times\frac{\left(\nabla_y\phi_\alpha\right)_{i,j+\frac{1}{2},k}}{\left|\nabla\phi_\alpha\right|_{i,j+1/2,k}}
$$

$$
\left(j_{at}^z\right)_{i,j,k+\frac{1}{2}} = \left[-\frac{\pi\varepsilon}{4}\frac{g_\alpha\left(\phi_\alpha\right)\left(1-h_\alpha\left(\phi_\alpha\right)\right)}{\sqrt{\phi_\alpha\left(1-\phi_\alpha\right)}}\left(c^\beta\left(\mu,T\right)-c^\alpha\left(\mu,T\right)\right)\frac{\partial\phi_\alpha}{\partial t}\right]_{i,j,k+\frac{1}{2}}
$$

$$
\times\frac{\left(\nabla_z\phi_\alpha\right)_{i,j,k+\frac{1}{2}}}{\left|\nabla\phi_\alpha\right|_{i,j,k+1/2}}.
$$

Next we elaborate each of the terms in the expressions.

$$
\left[-\frac{\pi\varepsilon}{4}\frac{g_\alpha\left(\phi_\alpha\right)\left(1-h_\alpha\left(\phi_\alpha\right)\right)}{\sqrt{\phi_\alpha\left(1-\phi_\alpha\right)}}\left(c^\beta\left(\mu,T\right)-c^\alpha\left(\mu,T\right)\right)\frac{\partial\phi_\alpha}{\partial t}\right]_{i+1/2,j,k} =
$$

$$
\frac{1}{2}\left[-\frac{\pi\varepsilon}{4}\frac{g_\alpha\left(\phi_\alpha\right)\left(1-h_\alpha\left(\phi_\alpha\right)\right)}{\sqrt{\phi_\alpha\left(1-\phi_\alpha\right)}}\left(c^\beta\left(\mu,T\right)-c^\alpha\left(\mu,T\right)\right)\frac{\partial\phi_\alpha}{\partial t}\right]_{i,j,k} +
$$

$$
\frac{1}{2}\left[-\frac{\pi\varepsilon}{4}\frac{g_\alpha\left(\phi_\alpha\right)\left(1-h_\alpha\left(\phi_\alpha\right)\right)}{\sqrt{\phi_\alpha\left(1-\phi_\alpha\right)}}\left(c^\beta\left(\mu,T\right)-c^\alpha\left(\mu,T\right)\right)\frac{\partial\phi_\alpha}{\partial t}\right]_{i+1,j,k}.
$$

$$
\left[-\frac{\pi\varepsilon}{4}\frac{g_\alpha\left(\phi_\alpha\right)\left(1-h_\alpha\left(\phi_\alpha\right)\right)}{\sqrt{\phi_\alpha\left(1-\phi_\alpha\right)}}\left(c^\beta\left(\mu,T\right)-c^\alpha\left(\mu,T\right)\right)\frac{\partial\phi_\alpha}{\partial t}\right]_{i,j+1/2,k} =
$$

$$\frac{1}{2}\left[-\frac{\pi\varepsilon}{4}\frac{g_\alpha\left(\phi_\alpha\right)\left(1-h_\alpha\left(\phi_\alpha\right)\right)}{\sqrt{\phi_\alpha\left(1-\phi_\alpha\right)}}\left(c^\beta\left(\mu,T\right)-c^\alpha\left(\mu,T\right)\right)\frac{\partial\phi_\alpha}{\partial t}\right]_{i,j,k}+$$

$$\frac{1}{2}\left[-\frac{\pi\varepsilon}{4}\frac{g_\alpha\left(\phi_\alpha\right)\left(1-h_\alpha\left(\phi_\alpha\right)\right)}{\sqrt{\phi_\alpha\left(1-\phi_\alpha\right)}}\left(c^\beta\left(\mu,T\right)-c^\alpha\left(\mu,T\right)\right)\frac{\partial\phi_\alpha}{\partial t}\right]_{i,j+1,k}.$$

$$\left[-\frac{\pi\varepsilon}{4}\frac{g_\alpha\left(\phi_\alpha\right)\left(1-h_\alpha\left(\phi_\alpha\right)\right)}{\sqrt{\phi_\alpha\left(1-\phi_\alpha\right)}}\left(c^\beta\left(\mu,T\right)-c^\alpha\left(\mu,T\right)\right)\frac{\partial\phi_\alpha}{\partial t}\right]_{i,j,k+1/2}=$$

$$\frac{1}{2}\left[-\frac{\pi\varepsilon}{4}\frac{g_\alpha\left(\phi_\alpha\right)\left(1-h_\alpha\left(\phi_\alpha\right)\right)}{\sqrt{\phi_\alpha\left(1-\phi_\alpha\right)}}\left(c^\beta\left(\mu,T\right)-c^\alpha\left(\mu,T\right)\right)\frac{\partial\phi_\alpha}{\partial t}\right]_{i,j,k}+$$

$$\frac{1}{2}\left[-\frac{\pi\varepsilon}{4}\frac{g_\alpha\left(\phi_\alpha\right)\left(1-h_\alpha\left(\phi_\alpha\right)\right)}{\sqrt{\phi_\alpha\left(1-\phi_\alpha\right)}}\left(c^\beta\left(\mu,T\right)-c^\alpha\left(\mu,T\right)\right)\frac{\partial\phi_\alpha}{\partial t}\right]_{i,j,k+1}.$$

The gradients are derived at the off-grid position as,

$$\left(\nabla_x\phi_\alpha\right)_{i+1/2,j,k}=\frac{\left(\phi_\alpha\right)_{i+1,j,k}-\left(\phi_\alpha\right)_{i,j,k}}{\Delta x}$$

$$\left(\nabla_y\phi_\alpha\right)_{i,j+1/2,k}=\frac{\left(\phi_\alpha\right)_{i,j+1,k}-\left(\phi_\alpha\right)_{i,j,k}}{\Delta y}$$

$$\left(\nabla_z\phi_\alpha\right)_{i,j,k+1/2}=\frac{\left(\phi_\alpha\right)_{i,j,k+1}-\left(\phi_\alpha\right)_{i,j,k}}{\Delta z}.$$

Finally we need to discretize the magnitude of the gradients at the respective off grid positions, which is done as follows,

$$\left|\nabla\phi_\alpha\right|_{(i+1/2,j,k)}=\sqrt{\left(\nabla_x\phi_\alpha\right)^2_{i+1/2,j,k}+\left(\nabla_y\phi_\alpha\right)^2_{i+1/2,j,k}+\left(\nabla_z\phi_\alpha\right)^2_{i+1/2,j,k}}$$

The gradient $\left(\nabla_x\phi_\alpha\right)_{i+1/2,j,k}$ at the position is already given before, while the gradients in the other directions are derived as follows,

$$\left(\nabla_y\phi_\alpha\right)_{i+1/2,j,k}=\frac{1}{2}\left(\left(\nabla_y\phi_\alpha\right)_{i,j,k}+\left(\nabla_y\phi_\alpha\right)_{i+1,j,k}\right)$$

$$\left(\nabla_y\phi_\alpha\right)_{i,j,k}=\frac{\left(\left(\phi_\alpha\right)_{i,j+1,k}-\left(\phi_\alpha\right)_{i,j-1,k}\right)}{2\Delta y}$$

$$(\nabla_y \phi_\alpha)_{i+1,j,k} = \frac{\left((\phi_\alpha)_{i+1,j+1,k} - (\phi_\alpha)_{i+1,j-1,k}\right)}{2\Delta y}$$

$$(\nabla_z \phi_\alpha)_{i+1/2,j,k} = \frac{1}{2}\left((\nabla_z \phi_\alpha)_{i,j,k} + (\nabla_z \phi_\alpha)_{i+1,j,k}\right)$$

$$(\nabla_z \phi_\alpha)_{i,j,k} = \frac{\left((\phi_\alpha)_{i,j,k+1} - (\phi_\alpha)_{i,j,k-1}\right)}{2\Delta z}$$

$$(\nabla_y \phi_\alpha)_{i+1,j,k} = \frac{\left((\phi_\alpha)_{i+1,j,k+1} - (\phi_\alpha)_{i+1,j,k-1}\right)}{2\Delta z}.$$

Similarly, the vector-norms at the other offset grid positions can be determined.

$$|\nabla \phi_\alpha|_{(i,j+1/2,k)} = \sqrt{(\nabla_x \phi_\alpha)^2_{i,j+1/2,k} + (\nabla_y \phi_\alpha)^2_{i,j+1/2,k} + (\nabla_z \phi_\alpha)^2_{i,j+1/2,k}}$$

The gradient $(\nabla_y \phi_\alpha)_{i,j+1/2,k}$ is given before. The other gradients are derived as,

$$(\nabla_x \phi_\alpha)_{i,j+1/2,k} = \frac{1}{2}\left((\nabla_x \phi_\alpha)_{i,j,k} + (\nabla_x \phi_\alpha)_{i,j+1,k}\right).$$

$$(\nabla_x \phi_\alpha)_{i,j,k} = \frac{\left((\phi_\alpha)_{i+1,j,k} - (\phi_\alpha)_{i-1,j,k}\right)}{2\Delta x}$$

$$(\nabla_x \phi_\alpha)_{i,j+1,k} = \frac{\left((\phi_\alpha)_{i+1,j+1,k} - (\phi_\alpha)_{i-1,j+1,k}\right)}{2\Delta x}$$

$$(\nabla_z \phi_\alpha)_{i,j+1/2,k} = \frac{1}{2}\left((\nabla_z \phi_\alpha)_{i,j,k} + (\nabla_z \phi_\alpha)_{i,j+1,k}\right).$$

$$(\nabla_z \phi_\alpha)_{i,j,k} = \frac{\left((\phi_\alpha)_{i,j,k+1} - (\phi_\alpha)_{i,j,k-1}\right)}{2\Delta z}$$

$$(\nabla_z \phi_\alpha)_{i,j+1,k} = \frac{\left((\phi_\alpha)_{i,j+1,k+1} - (\phi_\alpha)_{i,j+1,k-1}\right)}{2\Delta z}.$$

Finally, the norm on the offset grid position in the z-direction given by

$$|\nabla \phi_\alpha|_{(i,j,k+1/2)} = \sqrt{(\nabla_x \phi_\alpha)^2_{i,j,k+1/2} + (\nabla_y \phi_\alpha)^2_{i,j,k+1/2} + (\nabla_z \phi_\alpha)^2_{i,j,k+1/2}}.$$

The gradient in the z-direction $(\nabla_z \phi_\alpha)_{i,j,k+1/2}$ is given before while the gradient in the other directions can be derived as,

$$(\nabla_x \phi_\alpha)_{i,j,k+1/2} = \frac{1}{2}\left((\nabla_x \phi_\alpha)_{i,j,k} + (\nabla_x \phi_\alpha)_{i,j,k+1}\right)$$

$$(\nabla_x \phi_\alpha)_{i,j,k} = \frac{\left((\phi_\alpha)_{i+1,j,k} - (\phi_\alpha)_{i-1,j,k}\right)}{2\Delta x}$$

$$(\nabla_x \phi_\alpha)_{i,j,k+1} = \frac{\left((\phi_\alpha)_{i+1,j,k+1} - (\phi_\alpha)_{i-1,j,k+1}\right)}{2\Delta x}$$

$$(\nabla_y \phi_\alpha)_{i,j,k+1/2} = \frac{1}{2}\left((\nabla_y \phi_\alpha)_{i,j,k} + (\nabla_y \phi_\alpha)_{i,j,k+1}\right)$$

$$(\nabla_y \phi_\alpha)_{i,j,k} = \frac{\left((\phi_\alpha)_{i,j+1,k} - (\phi_\alpha)_{i,j-1,k}\right)}{2\Delta y}$$

$$(\nabla_y \phi_\alpha)_{i,j,k+1} = \frac{\left((\phi_\alpha)_{i,j+1,k+1} - (\phi_\alpha)_{i,j-1,k+1}\right)}{2\Delta y}.$$

## E.1. Parallelization

In this section, we discuss the terms transferred for parallelization in the z-direction. While all terms, related to the gradients in the phase-field variable $\phi$ are completely calculated with the imposition of a single boundary layer for each worker, which contains the information about the values of the $\phi_\alpha$ from the neighboring worker, we require additional transfer variables $\dfrac{\partial \phi_\alpha}{\partial t}$, for the anti-trapping current. We calculate this term by the following expression:

$$\frac{\partial \phi_\alpha}{\partial t} \approx \frac{\phi_\alpha^{n+1} - \phi_\alpha^n}{\Delta t}.$$

To understand, how this can be done correctly and effectively, it is essential to have a look at the calculation procedure in the domain. For the original (without the anti-trapping current), we use three buffer layers(gradient layers) for efficient utilization of the memory resources. This suffices since we use the nearest neighbor and second nearest neighbors for the dicretization of all the gradients. This is done as follows: For all points in a layer, we calculate one-sided gradients in the positive direction. Then the layers are swapped such that the next layer gradients are calculated. After this we have the relevant gradients in the $z-$ direction, that we need to compute the divergence as:

$$\nabla \cdot (\nabla \phi_\alpha)_{i,j,k} = \frac{\nabla_z^{i,j,k} \phi_\alpha - \nabla_z^{i,j,k-1} \phi_\alpha}{\Delta z},$$

where $\nabla_z^{i,j,k-1}$ contains the gradient in the earlier layer (z-1). With the divergence calculations, we are in a position to calculate all the terms required for the evolution of the $\phi_\alpha$ equation.

For the antitrapping term however, we need terms related to $\dfrac{\partial \phi_\alpha}{\partial t}$ at the off-grid positions. And we have seen in the discretization this is done through averaging the values of the cell of calculation and next neighbor cell in the direction. This requires that we have the $\phi_\alpha$ update of the present cell and the next cell in the positive direction before we calculate the update for the $c$ field. Hence, this requires that $c$ calculation be one step behind that of the phase-field such that all terms related to the $\phi_\alpha$ are from the time-step (n+1). The $\mu$ however is from the time step (n-1), which is consistent with the mass-conservation being satisfied at time-step (n-1). To achieve this whole thing entirely, requires that we increase the number of gradient layers by 1 and we structure our gradient layers naming them -1,0,1,2. While the calculation of $\phi_\alpha$ uses the gradient layers 0,1,2, the calculation of the concentration field is one step behind, and consequently one swap behind the phase-field and therefore comprises of -1,0,1. This now enables the computation in the parallelization direction to be of the same type as the other directions.

## Mobilities

The mobilities $M(\phi, c, T)$ are also required at the off-grid positions which are computed similarly by averaging the cell value and that of the neighbour cell. Since, the concentration is one iteration behind, we utilize the $\phi_\alpha^{n+1}$ for the computation of $M(T, c, \phi)$.

## MPI exchange

In the present parallelization scheme, we divide the domain in slices, parallel to the $z-$direction and the present dicretization scheme requires that we have the $\dfrac{\partial \phi_\alpha}{\partial t}$ at the boundary cells when we compute the antitrapping flux term for the boundary cells in the $z-$ direction. This is however impossible in the normal calculation scheme, since the update of the boundary cells is only after the complete calculation of the whole slice. To achieve this we do the following: For the the last boundary cell, we calculate the change in $\phi_\alpha$ from the neighboring worker, and store it in array for future calculation, before the start of the iteration for the entire slice. Note, this calculation does not modify the $\phi_\alpha$ values, but only computes the change in $\phi_\alpha$. For eg: For the cell $Nz - 1$, we compute the $\dfrac{\partial \phi_\alpha}{\partial t}$ of the layer $z = 1$ from the next worker, and similarly the change of $\phi_\alpha$ for $z = 0$ from $\dfrac{\partial \phi_\alpha}{\partial t}$ of the the layer $Nz - 2$ from the previous layer. The exchange of variables and the movement of the extended gradient layer can be seen in Figure E.1. This requires however, that we update the $\phi_\alpha$ values of the boundary before the $c$ calculation. This is performed through a manual update using the change in $\phi_\alpha$ values received from the neigboring workers. This update is essential both for the correct calculation of the mobilities and the correct calculation of the antitrapping current.

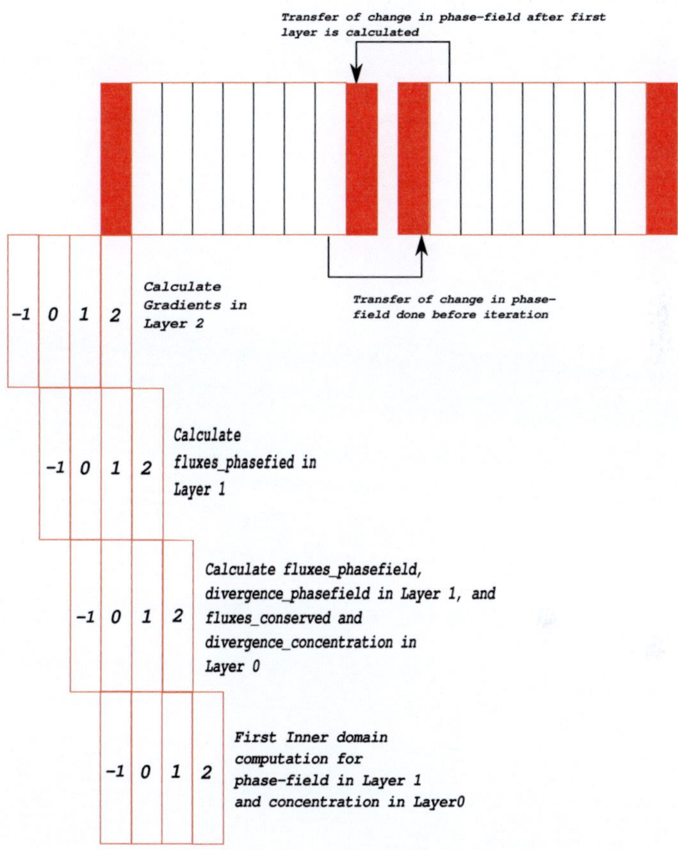



**Figure E.1.:** Modifications implemented, namely the increase of the gradient layers and the extra transfer variable "the change of the phase-field".

# Appendix F

# List of symbols

# F.1. List of symbols

| Symbol | Description | Units |
| --- | --- | --- |
| $\sigma_{\alpha\beta}, \gamma_{\alpha\beta}$ | Surface entropy | $J/m^2 K$ |
| $\tilde{\sigma}_{\alpha\beta}, \overline{\sigma_{\alpha\beta}}$ | Surface energy | $J/m^2$ |
| $\tau_{\alpha\beta}, \tau$ | Mobility(Phase-field(Free Energy Functional)) | $Js/m^4$ |
| $\omega_{\alpha\beta}, \omega$ | Mobility(Phase-field(Entropy Functional)) | $Js/m^4 K$ |
| $f$ | bulk free energy density | $J/m^3$ |
| $s$ | bulk entropy density | $J/m^3 K$ |
| $\Psi$ | grand potential density | $J/m^3$ |
| $\mathcal{S}$ | Entropy Functional | $J/K$ |
| $\mathcal{F}$ | Free Energy Functional | $J$ |
| $\Omega$ | Grand Potential Functional | $J$ |
| $\tilde{\tau}$ | Non-dimensional mobility(free energy) | - |
| $\zeta$ | Non-dimensional mobility(grand potential) | - |
| $\Lambda$ | Lagrange Parameter | $J/m^3 K$ |
| $\tilde{\Lambda}_{\alpha\beta}$ | Interface Thickness | $m$ |
| $M_{ij}$ | Mobilities(Concentration Equation) | $Km^3/J$ |
| $\tilde{M}, \tilde{F}$ | Solvability Integrals | - |
| $V_m$ | Molar Volume | $1/m^3$ |
| $R$ | Gas Constant | $J/K$ |
| $T$ | Temperature | $K$ |
| $u$ | Non-dimensional undercooling | - |
| $C_v$ | Heat capacity | $J/m^3 K$ |
| $K$ | Thermal Conductivity | $J/msK$ |
| $\kappa$ | Thermal Diffusivity | $m^2/s$ |
| $L_\alpha$ | Latent heat | $J/m^3$ |
| $\lambda$ | Lamellar spacing | $m$ |
| $\tilde{\lambda}$ | Temperature Scale | $K$ |
| $\tilde{\alpha}$ | Scaled Length(Diffusion Length-/Length scale) | - |

| Symbol | Description | Units |
|---|---|---|
| $p$ | Interface Peclet number | - |
| $g$ | Interpolation function, non-dimensional parameter | - |
| $h$ | Interpolation function | - |
| $D$ | Diffusivity | $m^2/s$ |
| $\Xi_\alpha$ | Noise Amplitude(Phase-field) | - |
| $r$ | Function of $\phi$ | - |
| $s$ | Function of $\phi$ | - |
| $\Gamma$ | Gibbs-Thomson Coefficient | $Km$ |
| $\mu$ | Chemical potential | $J/m^3$ |
| $\mu_{int}$ | Kinetic coefficient | $m/sK$ |
| $\bar{\beta}$ | $1/\mu_{int}$ | $Ks/m$ |
| $\mathcal{P}, \mathcal{Q}, \mathcal{R}, \mathcal{S}$ | Trignometric functions | - |
| $\alpha, \beta, \gamma, \delta$ | Phase indices | - |
| $i, j, k, A, B$ | Component Indices | - |
| $\varepsilon$ | Parameter relating to length(dimensional/non-dimensional) | $m/$- |
| $\boldsymbol{c}, \boldsymbol{\phi}$ | Phase and component vectors | - |
| $V$ | Velocity | $m/s$ |
| $v$ | Non-dimensional velocity | - |
| $l_c$ | Diffusion Length(D/V) | m |
| $d_o$ | Capillary length | m |
| $m$ | Slopes of Liquidus/Solius | $K$ |
| $t$ | time in simulations | $s$ |
| $k_n$ | Wave number | $1/m$ |
| $a$ | gradient entropy function | $J/m^2K$ |
| $\tilde{a}$ | gradient energy function | $J/m^2$ |
| $w$ | entropy potential | $J/m^2K$ |
| $\tilde{w}$ | energy potential | $J/m^2$ |

# Appendix G

# Acknowledgements

I would like to acknowledge the funding agencies: i)CCMSE (Center for Materials Science and Engineering) funded by the state of Baden-Wüttemberg, Germany and the European Fond for Regional Development (EFRE), ii)DFG(German Research Foundation-NE822/14-1), for the funding provided at the various stages of the PhD.

Many thanks to group members Sebastian Schulz, Marius Kist, Eugenia Wesner, Wang Fei, Cheng Qin, Oleg Tschukin, Daniel Schneider and Marouen Ben Said for involving me in various discussions on different topics, from which have generated new ideas.

I would like to acknowledge all the technical help provided by the computer scientists, Michael Selzer, Matthias Reichardt, Alexander Vondrous and Marcus Jainta at various stages during developments. Special gratitude for all the system administrators present and past who helped to solve problems simple and complex.

Special thanks to Marco Berghoff for generating the latex template for the dissertation.

Thanks to Frank Wendler, for helping to get started in the working group and for all the support and guidance.
Finally, special gratitude to Lorenz Ratke for providing the numerical data for the ternary eutectic alloy Al-Cu-Ag.

# Bibliography


[1] N. Ahmad, A. A. Wheeler, W. J. Boettinger, and G. B. McFadden. *Phys. Rev. E*, 58:3436, 1998.

[2] S. Akamatsu, S. Bottin-Rousseau, and G. Faivre. *Phys. Rev. Lett*, 93:175701, 2004.

[3] S. Akamatsu, S. Moulinet, and G. Faivre. *Met. Mat. Trans. A*, 32:2039, 2001.

[4] S. Akamatsu, M. Plapp, G. Faivre, and A. Karma. *Phys. Rev. E*, 66:030501, 2002.

[5] S. Akamatsu, M. Plapp, G. Faivre, and A. Karma. *Met. Mat. Trans. A*, 35:1815, 2004.

[6] R. F. Almgren. *SIAM J. Appl. Math*, 59:2086, 1999.

[7] M. Apel, B. Böttger, V. Witusiewicz, U. Hecht, and I. Steinbach. *Solidification and Crystallization, ed. by D. M. Herlach*. Weinheim: Wiley-VCH, 2004.

[8] D. Banerjee, R. Banerjee, and Y. Wang. *Scripta Materialia*, 41:1023, 1999.

[9] H. A. Quac Bao and F. C. L Durand. *J. Cryst. Growth*, 15:291, 1972.

[10] A. Barbieri and J. S. Langer. *Phys. Rev. A*, 39:5314, 1989.

[11] W. J. Boettinger and J. A. Warren. *Met. Mat. Trans. A*, 27A:657, 1996.

[12] M. H. Braga, J. C. R. E. Oliviera, L. F. Malheiros, and J. A. Ferreira. *CALPHAD:Computer Coupling of Phase Diagrams and Thermochemistry*, 33:237, 2009.

[13] G. Caginalp and W. Xie. *Phys. Rev. E*, 48:1897, 1993.

[14] J. W. Cahn and J. E. Hilliard. *J. Chem. Phys.*, 28:258, 1958.

[15] P. R. Cha, D. H. Yeon, and J. K. Yoon. *Acta. Mater.*, 49:3925, 2001.

[16] P. R. Cha, D. H. Yeon, and J. K. Yoon. *J. Cryst. Growth*, 274:281, 2005.

[17] Q. Chen, N. Ma, K. Wu, and Y. Wang. *Scripta Materialia*, 50:471, 2004.

[18] A. Choudhury and B. Nestler. *Phys.Rev.E*, 85:021602, 2011.

[19] A. Choudhury, B. Nestler, A. Telang, M. Selzer, and F. Wendler. *Acta Materialia*, 58:3815, 2010.

[20] A. Choudhury, M. Plapp, and B. Nestler. *Phys. Rev. E*, 83:051608, 2011.

[21] M. Conti. *Phys. Rev. E*, 55:765, 1997.

[22] M. Conti. *Phys. Rev. E*, 61:642, 2000.

[23] D. J. S. Cooksey and A. Hellawell. *J. Inst. Met.*, 95:183, 1967.

[24] P. Coullet and G. Iooss. *Phys. Rev. Lett.*, 64(8):866, 1990.

[25] D. Danilov and B. Nestler. *J. Cryst. Growth*, 275:e177, 2005.

[26] S. Dobler, T. S. Lo, M. Plapp, A. Karma, and W. Kurz. *Acta Mater.*, 52:2795, 2004.

[27] J. Eiken, B. Böttger, and I. Steinbach. *Phys. Rev. E*, 73:066122, 2006.

[28] K. R. Elder, M. Grant, N. Provatas, and J. M. Kosterlitz. *Phys. Rev. E*, 64:021604, 2001.

[29] H. Emmerich and R. Siqueri. *J. Phys. Condens. Matter.*, 18:11121, 2006.

[30] B. Eschebaria, R. Folch, A. Karma, and M. Plapp. *Phys. Rev. E*, 70:061604, 2004.

[31] R. Folch and M. Plapp. *Phys. Rev. E*, 68:010602, 2003.

[32] R. Folch and M. Plapp. *Phys. Rev. E*, 72:011602, 2005.

[33] H. Garcke, B. Nestler, and B. Stinner. *SIAM J Appl Math*, 64:775, 2004.

[34] M. Ginibre, S. Akamatsu, and G. Faivre. *Phys. Rev. E*, 56:780, 1997.

[35] V. L. Ginzburg and L. D. Landau. *Sov. Phys. JETP*, 20:1064, 1950.

[36] U. Grafe, B. Böttger, J. Tiaden, and S. G. Fries. *Scripta Mat.*, 42:1179, 2000.

[37] L. Gránásy, T. Börzsönyi, and T. Pusztai. *Phys. Rev. Lett.*, 88:201605, 2002.

[38] M. Gunduz and J. D. Hunt. *Acta. Metallurgica*, 33:1651, 1985.

[39] B. I. Halerpin and P. C. Hohenberg. *Phys. Rev. B*, 10:139, 1974.

[40] U. Hecht, L. Gránásy, T. Pusztai B. Böttger, M. Apel, V. Witusiewicz, L. Ratke, J. D. Wilde, L. Froyen, D. Camel, B. Drevet, G. Faivre, S. G. Fries, B. Legendre, and S. Rex. *Mat. Sci. Eng. R*, 46:1, 2004.

[41] T. Himemiya and T. Umeda. *Materials Transactions JIM*, 40(7):665–674, 1999.

[42] J. D. Holder and B. F. Oliver. *Mater. Trans.*, 5:2423, 1974.

[43] Y. L. Huang, A. Braachi, T. Niermann, M. Seibt, D. Danilov, B. Nestler, and S. Schneider. *Scripta Mat.*, 53:93, 2005.

[44] K. A. Jackson and J. D. Hunt. *Transaction of The Metallurgical Society of AIME*, 226:1129, 1966.

[45] A. Karma. *Phys. Rev. E*, 49:2245, 1993.

[46] A. Karma. *Phys. Rev. Lett.*, 87:115701, 2001.

[47] A. Karma and M. Plapp. *JOM*, 56:28, 2004.

[48] A. Karma and W. J. Rappel. *Phys. Rev. E*, 53(4):R3017, 1996.

[49] A. Karma and W. J. Rappel. *Journal of Crystal Growth*, 174:56–64, 1997.

[50] A. Karma and W. J. Rappel. *Phys. Rev. E*, 57:4323, 1998.

[51] A. Karma and A. Sarkissian. *Metall. Mater. Trans.*, 27A:635, 1996.

[52] H. W. Kerr, A. Plumtree, and W. C. Winegard. *J. Inst. Metals*, 93:63, 1964.

[53] K. B. Kim, J. Liu, N. Marasli, and J. D. Hunt. *Acta Metallurgica et Materialia*, 43:2143, 1995.

[54] S. G. Kim. *Acta Materialia*, 55:4391, 2007.

[55] S. G. Kim, W. T. Kim, and T. Suzuki. *Phys. Rev. E*, 58:3316, 1998.

[56] S. G. Kim, W. T. Kim, and T. Suzuki. *Phys. Rev. E*, 60:7186, 1999.

[57] S. G. Kim, W. T. Kim, and T. Suzuki. *J. Cryst. Growth*, 263:620, 2004.

[58] S. G. Kim, W. T. Kim, T. Suzuki, and M. Ode. *J. Cryst. Growth*, 261:135, 2004.

[59] Y. T. Kim, N. Provatas, N. Goldenfeld, and J. Dantzig. *Phys. Rev. E*, 59:2546, 1999.

[60] T. Kitashima and H. Harada. *Acta Materialia*, 57:2020, 2009.

[61] R. Kobayashi. *PhysicaD*, 63:410, 1993.

[62] J. S. Langer. *Phys. Rev. Lett.*, 44:1023, 1980.

[63] J. S. Langer. *Directions in Condensed Matter (World Scientific, Singapore)*, 165, 1986.

[64] J. S. Lee, S. G. Kim, W. T. Kim, and T. Suzuki. *ISIJ International*, 39(2):730, 1999.

[65] H. Levine. *Phys.Rev.B*, 31:6119, 1985.

[66] J. Lipton, M. E. Glicksman, and W. Kurz. *Mater. Sci. and Engg.*, 65:57, 1984.

[67] S. Liu, R. E. Napolitano, and R. Trivedi. *Acta. Materialia.*, 49:4271, 2001.

[68] T. S. Lo, S. Dobler, M. Plapp, A. Karma, and W. Kurz. *Acta Mater.*, 51:599, 2003.

[69] T. S. Lo, A. Karma, and M. Plapp. *Phys. Rev. E*, 63:031504, 2001.

[70] I. Loginova, J. Agren, and G. Amberg. *Acta Materialia*, 52:4055, 2004.

[71] I. Loginova, G. Amberg, and J. Ågren. *Acta. Mater.*, 49:573, 2001.

[72] W. Losert, A. Stillmann, H. Z. Cummins, Kopczyński, W. J. Rappel, and A. Karma. *Phys. Rev. E*, 58:7492, 1998.

[73] P. Mannevile. *Dissipative structures and weak turbulence*. Academic Press, Boston, 1990.

[74] D. G. McCartney, J. D. Hunt, and R. M. Jordan. *Met. Trans.*, 11A:1243, 1980.

[75] G. B. McFadden, A. A. Wheeler, and D. M. Anderson. *Physica D*, 144:154, 2000.

[76] Y. Mishin, W. J. Boettinger, J. A. Warren, and G. B. McFadden. *Acta Materialia*, 57:3771, 2009.

[77] B. Nestler and A. Choudhury. *Current Opinion in Solid State and Materials Science*, 15:93, 2011.

[78] B. Nestler, D. Danilov, A. Braachi, Y. L. Huang, T. Niermann, M. Seibt, and S. Schneider. *Mat. Sci. Engg. A*, 452:8, 2007.

[79] B. Nestler, H. Garcke, and B. Stinner. *Phys. Rev. E*, 71:041609, 2005.

[80] B. Nestler, F. Wendler, and M. Selzer. *Physical Review E*, 78:011604, 2008.

[81] B. Nestler and A. A. Wheeler. *PhysicaD*, 138:114, 2000.

[82] B. Nestler, A. A. Wheeler, L. Ratke, and C. Stöcker. *PhysicaD*, 141:133, 2000.

[83] M. Ohno and K. Matsuura. *Phys. Rev. E*, 79:031603, 2009.

[84] N. O. Opoku and N. Provatas. *Acta Materialia*, 58:2155, 2010.

[85] A. Parisi and M. Plapp. *Acta Mater.*, 56:1348, 2008.

[86] A. Parisi and M. Plapp. *EPL*, 90:26010, 2010.

[87] A. Parisi, M. Plapp, S. Akamatsu, S. Bottin-Rousseau, M. Perrut, and G. Faivre. Three-dimensional phase-field simulations of eutectic solidification and comparison to in-situ experimental observations, tms(the minerals, metals & materials society). 2000.

[88] O. Penrose and C. P. Fife. *PhysicaD*, 43:44, 1990.

[89] D. Phelan, M. Reid, and R. Dippenaar. *Mat. Sci. Engg. A*, 477:226, 2008.

[90] M. Plapp. *J. Cryst. Growth*, 303:49, 2007.

[91] M. Plapp. *Phys. Rev. E*, 84:031601, 2011.

[92] M. Plapp and A. Karma. *Phys. Rev. Lett.*, 84:1740, 2000.

[93] M. Plapp and A. Karma. *Phys. Rev. E*, 66:061608, 2002.

[94] N. Provatas, N. Goldenfeld, and J. Dantzig. *Phys. Rev. Lett.*, 80:3308, 1998.

[95] T. Pusztai, G. Tegze, G. I. T óth, L. Koernyei, G. Bansel, Z. Fan, and L. Gránásy. *J. Phys. Condens. Matter.*, 20:404205, 2008.

[96] R. S. Qin and E. R. Wallach. *Acta. Mater.*, 51:6199, 2003.

[97] D. Raabe and R. C. Becker. *Modeling Simul. Mater. Sci. Eng.*, 8:445, 2000.

[98] D. Raabe and L. Hantcher. *Comput. Mater. Sci.*, 34:299, 2005.

[99] J. C. Ramirez and C. Beckermann. Quantitative modeling of binary alloy solidification with coupled heat and solute diffusion via the phase-field method, aerospace sciences meeting reno, nevada 2003. 2003.

[100] L. Ratke. unpublished, 2010.

[101] S. Rex, B. Böttger, V. T. Witusiewicz, and U. Hecht. *J. Cryst. Growth*, 249:413–414, 2005.

[102] M. D. Rinaldi, R. M. Sharp, and M. C. Flemings. *Metall. Trans.*, 3:3139, 1972.

[103] G. Rubin and A. G. Khachaturiyan. *Acta Materialia*, 47(7):1995, 1999.

[104] M. A. Ruggiero and J. W. Rutter. *Mater. Sci. Technol.*, 11:136, 1995.

[105] H. Shibata, Y. Arai, M. Suzuki, and T. Emi. *Met. Mat. Trans. B*, 31B:981, 2000.

[106] J. P. Simmons, C. Shen, and Y. Wang. *Scripta Mat.*, 43:935, 2000.

[107] J. P. Simmons, Y. Wen, and Y. Z. Wang. *Materials Science and Engineering*, A365:136, 2004.

[108] I. Steinbach, F. Pezzola, B. Nestler, M. Seeselberg, R. Prieler, G. J. Schmitz, and J. L. L. Rezende. *PhysicaD*, 94:35, 1996.

[109] G. I. T óth and L. Gránásy. *J. Chem. Phys*, 127:074710, 2007.

[110] T. Takaki, Y. Hisakuni, T. Hirouchi, Y. Yamanaka, and Y. Tomita. *Comput. Mater. Sci.*, 45:881, 2009.

[111] T. Takaki and Y. Tomita. *Journal of Mechanical Sciences*, 52:320, 2010.

[112] G. Tegze, T. Pusztai, and L. Gránázy. *Mat. Sci. Engg. A*, 413:418, 2005.

[113] J. Tiaden. *Journal of Crystal Growth*, 198:1275, 1999.

[114] J. Tiaden, B. Nestler, H. J. Deipers, and I. Steinbach. *PhysicaD*, 115:73, 1998.

[115] G. I. Tóth and L. Gránásy. *J. Chem. Phys.*, 127:074709, 2007.

[116] R. Trivedi, J. T. Mason, J. D. Verhoeven, and W. Kurz. *Met. Trans. A*, 22:2523, 1991.

[117] V. Vaithyanathan and L. Q. Chen. *Acta Materialia*, 50:4061, 2002.

[118] H. Walker, S. Liu, J. H. Lee, and R. Trivedi. *Met. Mat. Trans. A*, 38A:1417, 2007.

[119] J. C. Wang, M. Osawa, T. Yokokawa, H. Harada, and M. Enomoto. *Computational Materials Science*, 39:871, 2007.

[120] J. C. Wang, M. Osawa, T. Yokokawa, H. Harada, and M. Enomoto. *Computational Materials Science*, 39:871, 2007.

[121] M. Wang, T. Jing, and B. Liu. *Scripta Mat.*, 61:777, 2009.

[122] M. Wang, B. Y. Zong, and G. Wang. *Computational Materials Science*, 45:217, 2009.

[123] S. L. Wang and R. F. Sekerka. *Phys. Rev. E*, 53:3760, 1996.

[124] S. L. Wang, R. F. Sekerka, A. A. Wheeler, B. T. Murray, S. R. Coriell, R. J. Braun, and J. B. McFadden. *PhysicaD*, 69:189, 1993.

[125] Y. Wang, D. Banerjee, C. C. Su, and A. G. Khachaturiyan. *Acta Materialia*, 46:2983, 1998.

[126] Y. Wang, L. Q. Chen, and A. G. Khachaturiyan. *Acta Metallurgica et. Materialia*, 41:279, 1993.

[127] Y. Wang and J. Li. *Acta Materialia*, 58:1212, 2010.

[128] J. A. Warren and W. Boettinger. *Acta. Metall. Mater.*, 43:689, 1995.

[129] J. A. Warren and W. J. Boettinger. *Acta. metall.*, 43(2):689, 1995.

[130] J. A. Warren and W. J. Boettinger. *Acta. Mater.*, 43(2):689, 1995.

[131] J. A. Warren, T. Pusztai, L. Környei, and L. Gránásy. *Phys. Rev. B*, 79:014204, 2009.

[132] Y. H. Wen, J. V. Lill, S. L. Chen, and J. P. Simmons. *Acta Materialia*, 58:875, 2010.

[133] Y. H. Wen, B. Wang, J. P. Simmons, and Y. Wang. *Mat. Transactions*, 54:2087, 2006.

[134] A. A. Wheeler, W. J. Boettinger, and G. B. McFadden. *Phys. Rev. E*, 45:7424, 1992.

[135] A. A. Wheeler, G. B. McFadden, and W. J. Boettinger. *Proc. R. Soc. Lond.*, 452:495, 1996.

[136] A. A. Wheeler, B. T. Murray, and R. J. Schaefer. *Physica D*, 66:243, 1993.

[137] V. T. Witusiewicz, U. Hecht, S. G. Fries, and S. Rex. *Journal of Alloys and Compounds*, 385:133, 2004.

[138] V. T. Witusiewicz, U. Hecht, S. G. Fries, and S. Rex. *Journal of Alloys and Compounds*, 387:217, 2005.

[139] V. T. Witusiewicz, U. Hecht, L. Sturz, and S. Rex. *J. Cryst. Growth*, 297:117, 2006.

[140] K. Wu, Y. A. Chang, and Y. Wang. *Scripta Materialia*, 50:1145, 2004.

[141] J. Z. Zhu, Z. K. Liu, V. Vaithyanathan, and L. Q. Chen. *Scripta Materialia*, 46:401, 2002.

## Schriftenreihe
## des Instituts für Angewandte Materialien

**ISSN 2192-9963**

Die Bände sind unter www.ksp.kit.edu als PDF frei verfügbar oder
als Druckausgabe bestellbar.

Band 1  Prachai Norajitra
**Divertor Development for a Future Fusion Power Plant.** 2011
ISBN 978-3-86644-738-7

Band 2  Jürgen Prokop
**Entwicklung von Spritzgießsonderverfahren zur Herstellung
von Mikrobauteilen durch galvanische Replikation.** 2011
ISBN 978-3-86644-755-4

Band 3  Theo Fett
**New contributions to R-curves and bridging stresses –
Applications of weight functions.** 2012
ISBN 978-3-86644-836-0

Band 4  Jérôme Acker
**Einfluss des Alkali/Niob-Verhältnisses und der Kupferdotierung
auf das Sinterverhalten, die Strukturbildung und die Mikro-
struktur von bleifreier Piezokeramik $(K_{0,5}Na_{0,5})NbO_3$.** 2012
ISBN 978-3-86644-867-4

Band 5  Holger Schwaab
**Nichtlineare Modellierung von Ferroelektrika unter
Berücksichtigung der elektrischen Leitfähigkeit.** 2012
ISBN 978-3-86644-869-8

Band 6  Christian Dethloff
**Modeling of Helium Bubble Nucleation and Growth
in Neutron Irradiated RAFM Steels.** 2012
ISBN 978-3-86644-901-5

Band 7  Jens Reiser
**Duktilisierung von Wolfram. Synthese, Analyse und Charak-
terisierung von Wolframlaminaten aus Wolframfolie.** 2012
ISBN 978-3-86644-902-2

Band 8  Andreas Sedlmayr
**Experimental Investigations of Deformation Pathways
in Nanowires.** 2012
ISBN 978-3-86644-905-3